Engineering Hydrology

Engineering Hydrology

Edited by
Wyatt Kelly

Larsen & Keller
www.larsen-keller.com

Engineering Hydrology
Edited by Wyatt Kelly
ISBN: 978-1-63549-105-0 (Hardback)

© 2017 Larsen & Keller

☰ Larsen & Keller

Published by Larsen and Keller Education,
5 Penn Plaza,
19th Floor,
New York, NY 10001, USA

Cataloging-in-Publication Data

Engineering hydrology / edited by Wyatt Kelly.
    p. cm.
Includes bibliographical references and index.
ISBN 978-1-63549-105-0
1. Hydraulic engineering. 2. Hydrology. I. Kelly, Wyatt.
TC145 .E54 2017
627--dc23

The publisher's policy is to use permanent paper from mills that operate a sustainable forestry policy. Furthermore, the publisher ensures that the text paper and cover boards used have met acceptable environmental accreditation standards.

Printed and bound in the United States of America.

For more information regarding Larsen and Keller Education and its products, please visit the publisher's website www.larsen-keller.com

Table of Contents

Preface

This book provides comprehensive insights into the field of hydrology. It traces the evolution of this field and talks in detail about its importance in the present day scenario. Hydrology refers to the study of the quality, distribution and movement of water and water bodies on Earth and other planets. It is also concerned with the study of environmental watershed sustainability, water resources and hydrologic cycle. The book highlights the fundamental components of engineering hydrology and its applicability with relation to concern of environmental conservation, food safety and the preservation of physical nature. It is a valuable compilation of topics, ranging from the basic to the most complex theories and principles in the field of engineering hydrology. Also included in it is a detailed explanation of the various concepts and applications of this subject. It aims to serve as a resource guide for students and facilitate the study of the discipline.

To facilitate a deeper understanding of the contents of this book a short introduction of every chapter is written below:

Chapter 1- Water is necessary to sustain life on Earth and though almost 80% of the Earth's surface is covered by water, it is a precious resource. Hydrology studies the water systems, changes in their structure, patterns of water distribution, their properties and the water or the hydrologic cycle. This chapter outlines the science of hydrology and helps the reader understand how water plays a pivotal role in many phenomena.

Chapter 2- Different branches of hydrology study the assorted issues related to water, its distribution, ecological importance, the sources of its replenishment etc. There are several branches of hydrology and some of them have been discussed in great depth with reference to the key concepts, principles, basic equations, methods and their applications. The chapter illustrates Ecohydrology, hydrogeology, hydroinformatics, hydrometeorology, isotope hydrology, limnology and surface-water hydrology.

Chapter 3- Fundamental features analyzed by hydrology include infiltration, routing, evapotranspiration and water quality. This chapter focuses on the characteristics, calculating methods and terminology of each feature. A section of the text discusses hydrograph and hydrography. This chapter provides a plethora of topics for better comprehension of hydrology.

Chapter 4- The tools used in hydrology measure features of rainwater and groundwater with respect to pressure, quality, heat transfer, infiltration, properties, etc. The tools considered closely include piezometer, infiltrometer and rain gauge while the techniques discussed are aquifer test, Darcy's law, Bowen's ratio, and watertable control. The text also examines the hydrological transport model.

Chapter 5- Hydrology finds valuable application in flood forecasting, hydroelectricity, antecedent moisture, drainage system, runoff model and discharge. This chapter considers how the tools and techniques of hydrology can be applied to improve human life. Agriculture is one of the major fields where hydrology is utilized.

Chapter 6- Hydraulic engineering studies the movement of fluids and its application in the design of bridges, water storage facilities, irrigation methods, dams, sewage systems, etc. Hydraulic engineering is involved in predicting the behavior of water and its interaction with these man-made structures. This chapter outlines hydraulic engineering and its importance to the field of civil engineering.

Chapter 7- Hydraulic engineering is principally employed in the construction of storm drains, levees and dams. It is used to study phenomenon like sediment transport and applying it to civil engineering structures related to water. This chapter also explores the uses and principles of the rain gauge, stream gauge, hygrometer and Green Kenue. The chapter strategically encompasses and incorporates the major components and key concepts of hydraulic engineering, providing a complete understanding.

Chapter 8- The science of hydrology incorporates numerous elements that study the ways and properties of geological water circulation and how water is continuously replenished naturally. This chapter discusses water cycle, water resources, water table, drainage basin, groundwater flow, aquifer, hydraulic conductivity, evaporation and moisture recycling. There is a section on water pollution as well.

Chapter 9- Floods occur when water inundates low lying areas which were not covered by water. They cause damage to both life and property. Floods can be caused due to the overflow of water bodies like lakes, rivers, streams etc. or by high precipitation. They can develop slowly or in a matter of minutes. This chapter comprehensively analyses floods, its types and related terminology like flood stage, flood warning, floodgate and flood barrier. The aspects elucidated in this chapter are of vital importance, and provide a better understanding of hydrology.

I owe the completion of this book to the never-ending support of my family, who supported me throughout the project.

Editor

Introduction to Hydrology

Water is necessary to sustain life on Earth and though almost 80% of the Earth's surface is covered by water, it is a precious resource. Hydrology studies the water systems, changes in their structure, patterns of water distribution, their properties and the water or the hydrologic cycle. This chapter outlines the science of hydrology and helps the reader understand how water plays a pivotal role in many phenomena.

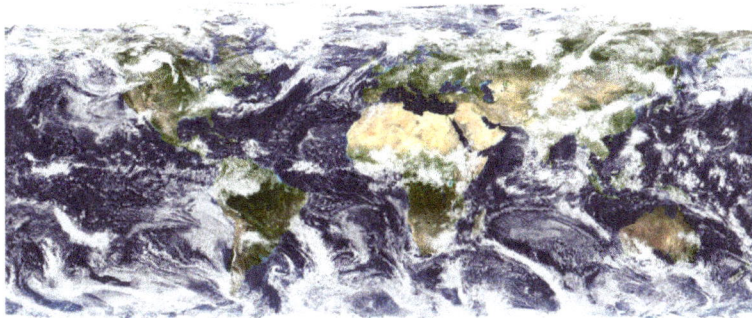

Water covers 70% of the Earth's surface.

Hydrology is the scientific study of the movement, distribution, and quality of water on Earth and other planets, including the hydrologic cycle, water resources and environmental watershed sustainability. A practitioner of hydrology is a hydrologist, working within the fields of earth or environmental science, physical geography, geology or civil and environmental engineering.

Hydrology subdivides into surface water hydrology, groundwater hydrology (hydrogeology), and marine hydrology. Domains of hydrology include hydrometeorology, surface hydrology, hydrogeology, drainage-basin management and water quality, where water plays the central role.

Oceanography and meteorology are not included because water is only one of many important aspects within those fields.

Hydrological research can inform environmental engineering, policy and planning.

The term *hydrology* comes from Greek: δωρ, *hydōr*, "water"; and λόγος, *logos*, "study".

History

Hydrology has been a subject of investigation and engineering for millennia. For example, about 4000 BC the Nile was dammed to improve agricultural productivity of previously barren lands. Mesopotamian towns were protected from flooding with high earthen walls. Aqueducts were built by the Greeks and Ancient Romans, while the history of China shows they built irrigation and flood control works. The ancient Sinhalese used hydrology to build complex irrigation works in Sri

Lanka, also known for invention of the Valve Pit which allowed construction of large reservoirs, anicuts and canals which still function.

Marcus Vitruvius, in the first century BC, described a philosophical theory of the hydrologic cycle, in which precipitation falling in the mountains infiltrated the Earth's surface and led to streams and springs in the lowlands. With adoption of a more scientific approach, Leonardo da Vinci and Bernard Palissy independently reached an accurate representation of the hydrologic cycle. It was not until the 17th century that hydrologic variables began to be quantified.

Pioneers of the modern science of hydrology include Pierre Perrault, Edme Mariotte and Edmund Halley. By measuring rainfall, runoff, and drainage area, Perrault showed that rainfall was sufficient to account for flow of the Seine. Marriotte combined velocity and river cross-section measurements to obtain discharge, again in the Seine. Halley showed that the evaporation from the Mediterranean Sea was sufficient to account for the outflow of rivers flowing into the sea.

Advances in the 18th century included the Bernoulli piezometer and Bernoulli's equation, by Daniel Bernoulli, and the Pitot tube, by Henri Pitot. The 19th century saw development in groundwater hydrology, including Darcy's law, the Dupuit-Thiem well formula, and Hagen-Poiseuille's capillary flow equation.

Rational analyses began to replace empiricism in the 20th century, while governmental agencies began their own hydrological research programs. Of particular importance were Leroy Sherman's unit hydrograph, the infiltration theory of Robert E. Horton, and C.V. Theis's aquifer test/equation describing well hydraulics.

Since the 1950s, hydrology has been approached with a more theoretical basis than in the past, facilitated by advances in the physical understanding of hydrological processes and by the advent of computers and especially geographic information systems (GIS).

Branches

- Chemical hydrology is the study of the chemical characteristics of water.

- Ecohydrology is the study of interactions between organisms and the hydrologic cycle.

- Hydrogeology is the study of the presence and movement of groundwater.

- Hydroinformatics is the adaptation of information technology to hydrology and water resources applications.

- Hydrometeorology is the study of the transfer of water and energy between land and water body surfaces and the lower atmosphere.

- Isotope hydrology is the study of the isotopic signatures of water.

- Surface hydrology is the study of hydrologic processes that operate at or near Earth's surface.

- Drainage basin management covers water-storage, in the form of reservoirs, and flood-protection.

- Water quality includes the chemistry of water in rivers and lakes, both of pollutants and natural solutes.

Applications

- Determining the water balance of a region.

- Determining the agricultural water balance.

- Designing riparian restoration projects.

- Mitigating and predicting flood, landslide and drought risk.

- Real-time flood forecasting and flood warning.

- Designing irrigation schemes and managing agricultural productivity.

- Part of the hazard module in catastrophe modeling.

- Providing drinking water.

- Designing dams for water supply or hydroelectric power generation.

- Designing bridges.

- Designing sewers and urban drainage system.

- Analyzing the impacts of antecedent moisture on sanitary sewer systems.

- Predicting geomorphologic changes, such as erosion or sedimentation.

- Assessing the impacts of natural and anthropogenic environmental change on water resources.

- Assessing contaminant transport risk and establishing environmental policy guidelines.

Themes

The central theme of hydrology is that water circulates throughout the Earth through different pathways and at different rates. The most vivid image of this is in the evaporation of water from the ocean, which forms clouds. These clouds drift over the land and produce rain. The rainwater flows into lakes, rivers, or aquifers. The water in lakes, rivers, and aquifers then either evaporates back to the atmosphere or eventually flows back to the ocean, completing a cycle. Water changes its state of being several times throughout this cycle.

The areas of research within hydrology concern the movement of water between its various states, or within a given state, or simply quantifying the amounts in these states in a given region. Parts of hydrology concern developing methods for directly measuring these flows or amounts of water, while others concern modelling these processes either for scientific knowledge or for making prediction in practical applications.

Groundwater

Ground water is water beneath Earth's surface, often pumped for drinking water. Groundwater hydrology (hydrogeology) considers quantifying groundwater flow and solute transport. Problems in describing the saturated zone include the characterization of aquifers in terms of flow direction, groundwater pressure and, by inference, groundwater depth. Measurements here can be made using a piezometer. Aquifers are also described in terms of hydraulic conductivity, storativity and transmisivity. There are a number of geophysical methods for characterising aquifers. There are also problems in characterising the vadose zone (unsaturated zone).

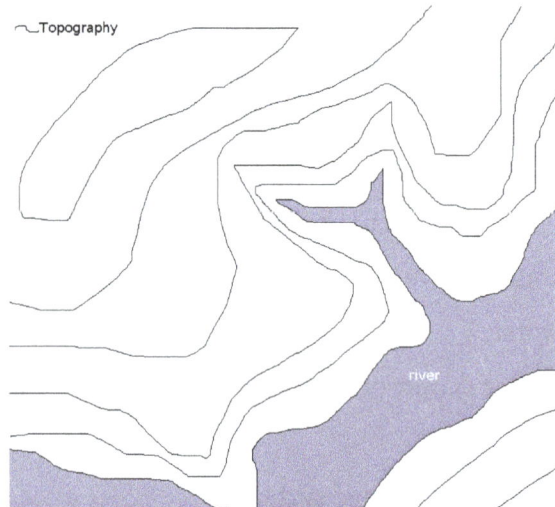
Building a map of groundwater contours

Infiltration

Infiltration is the process by which water enters the soil. Some of the water is absorbed, and the rest percolates down to the water table. The infiltration capacity, the maximum rate at which the soil can absorb water, depends on several factors. The layer that is already saturated provides a resistance that is proportional to its thickness, while that plus the depth of water above the soil provides the driving force (hydraulic head). Dry soil can allow rapid infiltration by capillary action; this force diminishes as the soil becomes wet. Compaction reduces the porosity and the pore sizes. Surface cover increases capacity by retarding runoff, reducing compaction and other processes. Higher temperatures reduce viscosity, increasing infiltration.

Soil Moisture

Soil moisture can be measured in various ways; by capacitance probe, time domain reflectometer or Tensiometer. Other methods include solute sampling and geophysical methods.

Surface Water Flow

Hydrology considers quantifying surface water flow and solute transport, although the treatment of flows in large rivers is sometimes considered as a distinct topic of hydraulics or hydrodynamics. Surface water flow can include flow both in recognizable river channels and otherwise. Methods for measuring flow once water has reached a river include the stream gauge, and tracer techniques. Other topics include chemical transport as part of surface water, sediment transport and erosion.

One of the important areas of hydrology is the interchange between rivers and aquifers. Ground-water/surface water interactions in streams and aquifers can be complex and the direction of net water flux (into surface water or into the aquifer) may vary spatially along a stream channel and over time at any particular location, depending on the relationship between stream stage and groundwater levels.

Precipitation and Evaporation

In some considerations, hydrology is thought of as starting at the land-atmosphere boundary and so it is important to have adequate knowledge of both precipitation and evaporation. Precipitation can be measured in various ways: disdrometer for precipitation characteristics at a fine time scale; radar for cloud properties, rain rate estimation, hail and snow detection; Rain gauge for routine accurate measurements of rain and snowfall; satellite – rainy area identification, rain rate estimation, land-cover/land-use, soil moisture.

Evaporation is an important part of the water cycle. It is partly affected by humidity, which can be measured by a sling psychrometer. It is also affected by the presence of snow, hail and ice and can relate to dew, mist and fog. Hydrology considers evaporation of various forms: from water surfaces; as transpiration from plant surfaces in natural and agronomic ecosystems. A direct measurement of evaporation can be obtained using Symon's evaporation pan.

Detailed studies of evaporation involve boundary layer considerations as well as momentum, heat flux and energy budgets.

Remote Sensing

Remote sensing of hydrologic processes can provide information of various types. Sources include land based sensors, airborne sensors and satellite sensors. Information can include clouds, surface moisture, vegetation cover.

Water Quality

In hydrology, studies of water quality concern organic and inorganic compounds, and both dissolved and sediment material. In addition, water quality is affected by the interaction of dissolved oxygen with organic material and various chemical transformations that may take place. Measurements of water quality may involve either in-situ methods, in which analyses take place on-site, often automatically, and laboratory-based analyses and may include microbiological analysis.

Integrating Measurement and Modelling

- Budget analyses
- Parameter estimation
- Scaling in time and space
- Data assimilation
- Quality control of data

Prediction

Observations of hydrologic processes are used to make predictions of the future behaviour of hydrologic systems (water flow, water quality). One of the major current concerns in hydrologic research is "Prediction in Ungauged Basins" (PUB), i.e. in basins where no or only very few data exist.

Statistical Hydrology

By analyzing the statistical properties of hydrologic records, such as rainfall or river flow, hydrologists can estimate future hydrologic phenomena. When making assessments of how often relatively rare events will occur, analyses are made in terms of the return period of such events. Other quantities of interest include the average flow in a river, in a year or by season.

These estimates are important for engineers and economists so that proper risk analysis can be performed to influence investment decisions in future infrastructure and to determine the yield reliability characteristics of water supply systems. Statistical information is utilized to formulate operating rules for large dams forming part of systems which include agricultural, industrial and residential demands.

Modeling

Hydrological models are simplified, conceptual representations of a part of the hydrologic cycle. They are primarily used for hydrological prediction and for understanding hydrological processes, within the general field of scientific modeling. Two major types of hydrological models can be distinguished:

- Models based on data. These models are black box systems, using mathematical and statistical concepts to link a certain input (for instance rainfall) to the model output (for instance runoff). Commonly used techniques are regression, transfer functions, and system identification. The simplest of these models may be linear models, but it is common to deploy non-linear components to represent some general aspects of a catchment's response without going deeply into the real physical processes involved. An example of such an aspect is the well-known behavior that a catchment will respond much more quickly and strongly when it is already wet than when it is dry..

- Models based on process descriptions. These models try to represent the physical processes observed in the real world. Typically, such models contain representations of surface runoff, subsurface flow, evapotranspiration, and channel flow, but they can be far more complicated. These models are known as deterministic hydrology models. Deterministic hydrology models can be subdivided into single-event models and continuous simulation models.

Recent research in hydrological modeling tries to have a more global approach to the understanding of the behavior of hydrologic systems to make better predictions and to face the major challenges in water resources management.

Transport

Water movement is a significant means by which other material, such as soil, gravel, boulders or pollutants, are transported from place to place. Initial input to receiving waters may arise from a point source discharge or a line source or area source, such as surface runoff. Since the 1960s rather complex mathematical models have been developed, facilitated by the availability of high speed computers. The most common pollutant classes analyzed are nutrients, pesticides, total dissolved solids and sediment.

References

- Vereecken, H.; Kemna, A.; Münch, H. M.; Tillmann, A.; Verweerd, A. (2006). "Aquifer Characterization by Geophysical Methods". Encyclopedia of Hydrological Sciences. John Wiley & Sons. doi:10.1002/0470848944. hsa154b. ISBN 0-471-49103-9.

- Wilson, L. Gray; Everett, Lorne G.; Cullen, Stephen J. (1994). Handbook of Vadose Zone Characterization & Monitoring. CRC Press. ISBN 978-0-87371-610-9.

- Reddy, P. Jaya Rami (2007). A textbook of hydrology (Reprint. ed.). New Delhi: Laxmi Publ. ISBN 9788170080992.

- "Eawag aquatic research". Swiss Federal Institute of Aquatic Science and Technology. 25 January 2012. Retrieved 8 March 2013.

- "National Hydrology Research Centre (Saskatoon, SK)". Environmental Science Centres. Environment Canada. Retrieved 8 March 2013.

- Гидрологическая комиссия [Hydrological Commission] (in Russian). Russian Geographical Society. Retrieved 8 March 2013.

Branches of Hydrology

Different branches of hydrology study the assorted issues related to water, its distribution, ecological importance, the sources of its replenishment etc. There are several branches of hydrology and some of them have been discussed in great depth with reference to the key concepts, principles, basic equations, methods and their applications. The chapter illustrates ecohydrology, hydrogeology, hydroinformatics, hydrometeorology, isotope hydrology, limnology and surface-water hydrology.

Ecohydrology

Ecohydrology is an interdisciplinary field studying the interactions between water and ecosystems. These interactions may take place within water bodies, such as rivers and lakes, or on land, in forests, deserts, and other terrestrial ecosystems. Areas of research in ecohydrology include transpiration and plant water use, adaption of organisms to their water environment, influence of vegetation on stream flow and function, and feedbacks between ecological processes and the hydrological cycle.

Key Concepts

The hydrologic cycle describes the continuous movement of water on, above, and below the surface on the earth. This flow is altered by ecosystems at numerous points. Transpiration from plants provides the majority of flow of water to the atmosphere. Water is influenced by vegetative cover as it flows over the land surface, while river channels can be shaped by the vegetation within them.

Ecohydrologists study both terrestrial and aquatic systems. In terrestrial ecosystems (such as forests, deserts, and savannas), the interactions among vegetation, the land surface, the vadose zone, and the groundwater are the main focus. In aquatic ecosystems (such as rivers, streams, lakes, and wetlands), emphasis is placed on how water chemistry, geomorphology, and hydrology affect their structure and function.

Principles

The principles of Ecohydrology are expressed in three sequential components:

1. Hydrological: The quantification of the hydrological cycle of a basin, should be a template for functional integration of hydrological and biological processes.

2. Ecological: The integrated processes at river basin scale can be steered in such a way as to enhance the basin's carrying capacity and its ecosystem services.

3. Ecological engineering: The regulation of hydrological and ecological processes, based on an integrative system approach, is thus a new tool for Integrated Water Basin Management.

Their expression as testable hypotheses (Zalewski et al., 1997) may be seen as:

-   H1: Hydrological processes generally regulate biota

-   H2: Biota can be shaped as a tool to regulate hydrological processes

-   H3: These two types of regulations (H1&H2) can be integrated with hydro-technical infrastructure to achieve sustainable water and ecosystem services

Vegetation and Water Stress

A fundamental concept in ecohydrology is that plant physiology is directly linked to water availability. Where there is ample water, as in rainforests, plant growth is more dependent on nutrient availability. However, in semi-arid areas, like African savannas, vegetation type and distribution relate directly to the amount of water that plants can extract from the soil. When insufficient soil water is available, a water-stressed condition occurs. Plants under water stress decrease both their transpiration and photosynthesis through a number of responses, including closing their stomata. This decrease in the canopy water flux and carbon dioxide flux can influence surrounding climate and weather.

Insufficient soil moisture produces stress in plants, and water availability is one of the two most important factors (temperature being the other) that determine species distribution. High winds, low atmospheric relative humidity, low carbon dioxide, high temperature, and high irradiance all exacerbate soil moisture insufficiency. Soil moisture availability is also reduced at low soil temperature. One of the earliest responses to insufficient moisture supply is a reduction in turgor pressure; cell expansion and growth are immediately inhibited, and unsuberized shoots soon wilt.

The concept of water deficit, as developed by Stocker in the 1920s, is a useful index of the balance in the plant between uptake and loss of water. Slight water deficits are normal and do not impair the functioning of the plant, while greater deficits disrupt normal plant processes.

An increase in moisture stress in the rooting medium as small as 5 atmospheres affects growth, transpiration, and internal water balance in seedlings, much more so in Norway spruce than in birch, aspen, or Scots pine. The decrease in net assimilation rate is greater in the spruce than in the other species, and, of those species, only the spruce shows no increase in water use efficiency as the soil becomes drier. The two conifers show larger differences in water potential between leaf and substrate than do the hardwoods. Transpiration rate decrease less in Norway spruce than in the other three species as soil water stress increases up to 5 atmospheres in controlled environments. In field conditions, Norway spruce needles lose three times as much water from the fully turgid state as do birch and aspen leaves, and twice as much as Scots pine, before apparent closure of stomata (although there is some difficulty in determining the exact point of closure). Assimilation may therefore continue longer in spruce than in pine when plant water stresses are high, though spruce will probably be the first to "run out of water".

Soil Moisture Dynamics

Soil moisture is a general term describing the amount of water present in the vadose zone, or unsaturated portion of soil below ground. Since plants depend on this water to carry out critical biological processes, soil moisture is integral to the study of ecohydrology. Soil moisture is generally described as water content, θ, or saturation, S. These terms are related by porosity, n, through the equation $\theta = nS$. The changes in soil moisture over time are known as soil moisture dynamics.

Recent global studies using water stable isotopes show that not all soil moisture is equally available for groundwater recharge or for plant transpiration.

Temporal and Spatial Considerations

Ecohydrological theory also places importance on considerations of temporal (time) and spatial (space) relationships. Hydrology, in particular the timing of precipitation events, can be a critical factor in the way an ecosystem evolves over time. For instance, Mediterranean landscapes experience dry summers and wet winters. If the vegetation has a summer growing season, it often experiences water stress, even though the total precipitation throughout the year may be moderate. Ecosystems in these regions have typically evolved to support high water demand grasses in the winter, when water availability is high, and drought-adapted trees in the summer, when it is low.

Ecohydrology also concerns itself with the hydrological factors behind the spatial distribution of plants. The optimal spacing and spatial organization of plants is at least partially determined by water availability. In ecosystems with low soil moisture, trees are typically located further apart than they would be in well-watered areas.

Basic Equations and Models

Water Balance at a Point

A fundamental equation in ecohydrology is the water balance at a point in the landscape. A water balance states that the amount water entering the soil must be equal to the amount of water leaving the soil plus the change in the amount of water stored in the soil. The water balance has four main components: infiltration of precipitation into the soil, evapotranspiration, leakage of water into deeper portions of the soil not accessible to the plant, and runoff from the ground surface. It is described by the following equation:

$$nZ_r \frac{ds(t)}{dt} = R(t) - I(t) - Q[s(t),t] - E[s(t)] - L[s(t)]$$

The terms on the left hand side of the equation describe the total amount of water contained in the rooting zone. This water, accessible to vegetation, has a volume equal to the porosity of the soil (n) multiplied by its saturation (s) and the depth of the plant's roots (Z_r). The differential equation $ds(t)/dt$ describes how the soil saturation changes over time. The terms on the right hand side describe the rates of rainfall (R), interception (I), runoff (Q), evapotranspiration (E), and leakage (L). These are typically given in millimeters per day (mm/d). Runoff, evaporation, and leakage are all highly dependent on the soil saturation at a given time.

In order to solve the equation, the rate of evapotranspiration as a function of soil moisture must be known. The model generally used to describe it states that above a certain saturation, evaporation will only be dependent on climate factors such as available sunlight. Once below this point, soil moisture imposes controls on evapotranspiration, and it decreases until the soil reaches the point where the vegetation can no longer extract any more water. This soil level is generally referred to as the "permanent wilting point". This term is confusing because many plant species do not actually "wilt".

Hydrogeology

Hydrogeology (*hydro-* meaning water, and *-geology* meaning the study of the Earth) is the area of geology that deals with the distribution and movement of groundwater in the soil and rocks of the Earth's crust (commonly in aquifers). The term geohydrology is often used interchangeably. Some make the minor distinction between a hydrologist or engineer applying themselves to geology (geohydrology), and a geologist applying themselves to hydrology (hydrogeology).

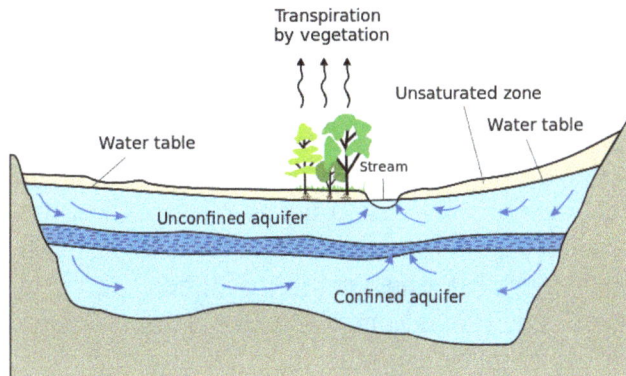

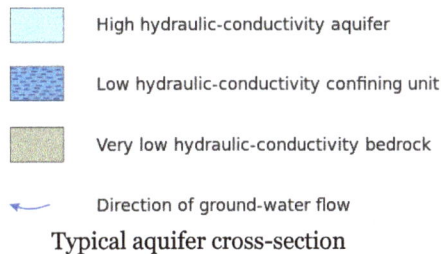

Typical aquifer cross-section

Introduction

Hydrogeology is an interdisciplinary subject; it can be difficult to account fully for the chemical, physical, biological and even legal interactions between soil, water, nature and society. The study of the interaction between groundwater movement and geology can be quite complex. Groundwater does not always flow in the subsurface down-hill following the surface topography; groundwater follows pressure gradients (flow from high pressure to low) often following fractures and conduits in circuitous paths. Taking into account the interplay of the different facets of a multi-component system often requires knowledge in several diverse fields at both the experimental and theoretical levels. The following is a more traditional introduction to the methods and nomenclature of saturated subsurface hydrology, or simply the study of ground water content.

Hydrogeology in Relation to Other Fields

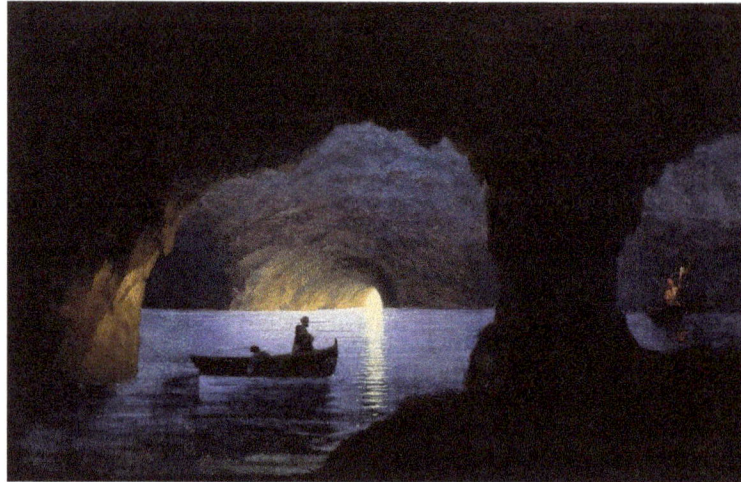

Painting of Ivan Aivazovsky (1841).

Hydrogeology, as stated above, is a branch of the earth sciences dealing with the flow of water through aquifers and other shallow porous media (typically less than 450 m or 1,500 ft below the land surface.) The very shallow flow of water in the subsurface (the upper 3 m or 10 ft) is pertinent to the fields of soil science, agriculture and civil engineering, as well as to hydrogeology. The general flow of fluids (water, hydrocarbons, geothermal fluids, etc.) in deeper formations is also a concern of geologists, geophysicists and petroleum geologists. Groundwater is a slow-moving, viscous fluid (with a Reynolds number less than unity); many of the empirically derived laws of groundwater flow can be alternately derived in fluid mechanics from the special case of Stokes flow (viscosity and pressure terms, but no inertial term).

The mathematical relationships used to describe the flow of water through porous media are the diffusion and Laplace equations, which have applications in many diverse fields. Steady groundwater flow (Laplace equation) has been simulated using electrical, elastic and heat conduction analogies. Transient groundwater flow is analogous to the diffusion of heat in a solid, therefore some solutions to hydrological problems have been adapted from heat transfer literature.

Traditionally, the movement of groundwater has been studied separately from surface water, climatology, and even the chemical and microbiological aspects of hydrogeology (the processes are uncoupled). As the field of hydrogeology matures, the strong interactions between groundwater, surface water, water chemistry, soil moisture and even climate are becoming more clear.

For example: Aquifer drawdown or overdrafting and the pumping of fossil water may be a contributing factor to sea-level rise.

Definitions and Material Properties

One of the main tasks a hydrogeologist typically performs is the prediction of future behavior of an aquifer system, based on analysis of past and present observations. Some hypothetical, but characteristic questions asked would be:

- Can the aquifer support another subdivision?

- Will the river dry up if the farmer doubles his irrigation?

- Did the chemicals from the dry cleaning facility travel through the aquifer to my well and make me sick?

- Will the plume of effluent leaving my neighbor's septic system flow to my drinking water well?

Most of these questions can be addressed through simulation of the hydrologic system (using numerical models or analytic equations). Accurate simulation of the aquifer system requires knowledge of the aquifer properties and boundary conditions. Therefore, a common task of the hydrogeologist is determining aquifer properties using aquifer tests.

In order to further characterize aquifers and aquitards some primary and derived physical properties are introduced below. Aquifers are broadly classified as being either confined or unconfined (water table aquifers), and either saturated or unsaturated; the type of aquifer affects what properties control the flow of water in that medium (e.g., the release of water from storage for confined aquifers is related to the storativity, while it is related to the specific yield for unconfined aquifers).

Hydraulic Head

Differences in hydraulic head (h) cause water to move from one place to another; water flows from locations of high h to locations of low h. Hydraulic head is composed of pressure head (ψ) and elevation head (z). The head gradient is the change in hydraulic head per length of flowpath, and appears in Darcy's law as being proportional to the discharge.

Hydraulic head is a directly measurable property that can take on any value (because of the arbitrary datum involved in the z term); ψ can be measured with a pressure transducer (this value can be negative, e.g., suction, but is positive in saturated aquifers), and z can be measured relative to a surveyed datum (typically the top of the well casing). Commonly, in wells tapping unconfined aquifers the water level in a well is used as a proxy for hydraulic head, assuming there is no vertical gradient of pressure. Often only *changes* in hydraulic head through time are needed, so the constant elevation head term can be left out ($\Delta h = \Delta \psi$).

A record of hydraulic head through time at a well is a hydrograph or, the changes in hydraulic head recorded during the pumping of a well in a test are called drawdown.

Porosity

Porosity (n) is a directly measurable aquifer property; it is a fraction between 0 and 1 indicating the amount of pore space between unconsolidated soil particles or within a fractured rock. Typically, the majority of groundwater (and anything dissolved in it) moves through the porosity available to flow (sometimes called effective porosity). Permeability is an expression of the connectedness of the pores. For instance, an unfractured rock unit may have a high *porosity* (it has lots of *holes* between its constituent grains), but a low *permeability* (none of the pores are connected). An example of this phenomenon is pumice, which, when in its unfractured state, can make a poor aquifer.

Porosity does not directly affect the distribution of hydraulic head in an aquifer, but it has a very strong effect on the migration of dissolved contaminants, since it affects groundwater flow velocities through an inversely proportional relationship.

Water Content

Water content (θ) is also a directly measurable property; it is the fraction of the total rock which is filled with liquid water. This is also a fraction between 0 and 1, but it must also be less than or equal to the total porosity.

The water content is very important in vadose zone hydrology, where the hydraulic conductivity is a strongly nonlinear function of water content; this complicates the solution of the unsaturated groundwater flow equation.

Hydraulic Conductivity

Hydraulic conductivity (K) and transmissivity (T) are indirect aquifer properties (they cannot be measured directly). T is the K integrated over the vertical thickness (b) of the aquifer ($T=Kb$ when K is constant over the entire thickness). These properties are measures of an aquifer's ability to transmit water. Intrinsic permeability (κ) is a secondary medium property which does not depend on the viscosity and density of the fluid (K and T are specific to water); it is used more in the petroleum industry.

Specific Storage and Specific Yield

Specific storage (S_s) and its depth-integrated equivalent, storativity ($S=S_s b$), are indirect aquifer properties (they cannot be measured directly); they indicate the amount of groundwater released from storage due to a unit depressurization of a confined aquifer. They are fractions between 0 and 1.

Specific yield (S_y) is also a ratio between 0 and 1 ($S_y \leq$ porosity) and indicates the amount of water released due to drainage from lowering the water table in an unconfined aquifer. The value for specific yield is less than the value for porosity because some water will remain in the medium even after drainage due to intermolecular forces. Often the porosity or effective porosity is used as an upper bound to the specific yield. Typically S_y is orders of magnitude larger than S_s.

Contaminant Transport Properties

Often we are interested in how the moving groundwater will transport dissolved contaminants around (the sub-field of contaminant hydrogeology). The contaminants can be man-made (e.g., petroleum products, nitrate, Chromium or radionuclides) or naturally occurring (e.g., arsenic, salinity). Besides needing to understand where the groundwater is flowing, based on the other hydrologic properties discussed above, there are additional aquifer properties which affect how dissolved contaminants move with groundwater.

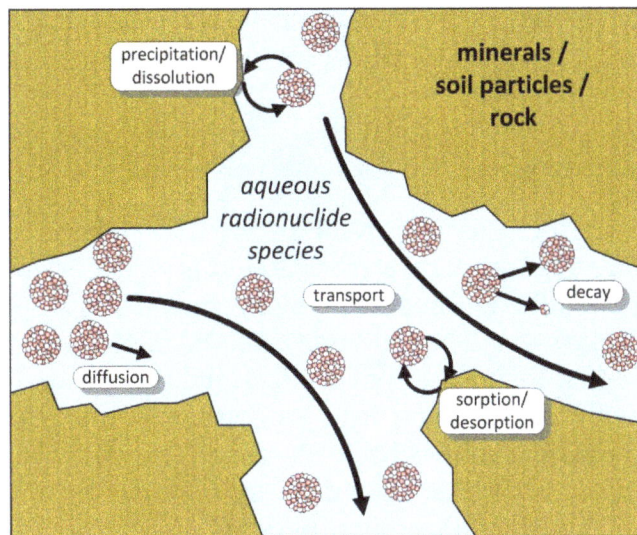

Transport and fate of contaminants in groundwater

Hydrodynamic Dispersion

Hydrodynamic dispersivity (α_L, α_T) is an empirical factor which quantifies how much contaminants stray away from the path of the groundwater which is carrying it. Some of the contaminants will be "behind" or "ahead" the mean groundwater, giving rise to a longitudinal dispersivity (α_L), and some will be "to the sides of" the pure advective groundwater flow, leading to a transverse dispersivity (α_T). Dispersion in groundwater arises because each water "particle", passing beyond a soil particle, must choose where to go, whether left or right or up or down, so that the water "particles" (and their solute) are gradually spread in all directions around the mean path. This is the "microscopic" mechanism, on the scale of soil particles. More important, on long distances, can be the macroscopic inhomogeneities of the aquifer, which can have regions of larger or smaller permeability, so that some water can find a preferential path in one direction, some other in a different direction, so that the contaminant can be spread in a completely irregular way, like in a (three-dimensional) delta of a river.

Dispersivity is actually a factor which represents our *lack of information* about the system we are simulating. There are many small details about the aquifer which are being averaged when using a macroscopic approach (e.g., tiny beds of gravel and clay in sand aquifers), they manifest themselves as an *apparent* dispersivity. Because of this, α is often claimed to be dependent on the length scale of the problem — the dispersivity found for transport through 1 m³ of aquifer is different from that for transport through 1 cm³ of the same aquifer material.

Molecular Diffusion

Diffusion is a fundamental physical phenomenon, which Einstein characterized as Brownian motion, that describes the random thermal movement of molecules and small particles in gases and liquids. It is an important phenomenon for small distances (it is essential for the achievement of thermodynamic equilibria), but, as the time necessary to cover a distance by diffusion is proportional to the square of the distance itself, it is ineffective for spreading a solute over macroscopic distances. The diffusion coefficient, D, is typically quite small, and its effect can often be considered negligible (unless groundwater flow velocities are extremely low, as they are in clay aquitards).

It is important not to confuse diffusion with dispersion, as the former is a physical phenomenon and the latter is an empirical factor which is cast into a similar form as diffusion, because we already know how to solve that problem.

Retardation By Adsorption

The retardation factor is another very important feature that make the motion of the contaminant to deviate from the average groundwater motion. It is analogous to the retardation factor of chromatography. Unlike diffusion and dispersion, which simply spread the contaminant, the retardation factor changes its *global average velocity*, so that it can be much slower than that of water. This is due to a chemico-physical effect: the adsorption to the soil, which holds the contaminant back and does not allow it to progress until the quantity corresponding to the chemical adsorption equilibrium has been adsorbed. This effect is particularly important for less soluble contaminants, which thus can move even hundreds or thousands times slower than water. The effect of this phenomenon is that only more soluble species can cover long distances. The retardation factor depends on the chemical nature of both the contaminant and the aquifer.

Governing Equations

Darcy's Law

Darcy's law is a constitutive equation, empirically derived by Henry Darcy in 1856, which states that the amount of groundwater discharging through a given portion of aquifer is proportional to the cross-sectional area of flow, the hydraulic gradient, and the hydraulic conductivity.

Groundwater Flow Equation

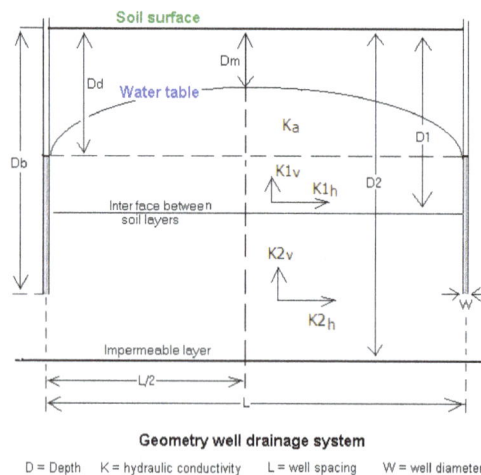

Geometry of a partially penetrating well drainage system in an anisotropic layered aquifer

The groundwater flow equation, in its most general form, describes the movement of groundwater in a porous medium (aquifers and aquitards). It is known in mathematics as the diffusion equation, and has many analogs in other fields. Many solutions for groundwater flow problems were borrowed or adapted from existing heat transfer solutions.

It is often derived from a physical basis using Darcy's law and a conservation of mass for a small control volume. The equation is often used to predict flow to wells, which have radial symmetry, so the flow equation is commonly solved in polar or cylindrical coordinates.

The Theis equation is one of the most commonly used and fundamental solutions to the groundwater flow equation; it can be used to predict the transient evolution of head due to the effects of pumping one or a number of pumping wells.

The Thiem equation is a solution to the steady state groundwater flow equation (Laplace's Equation) for flow to a well. Unless there are large sources of water nearby (a river or lake), true steady-state is rarely achieved in reality.

Both above equations are used in aquifer tests (pump tests).

The Hooghoudt equation is a groundwater flow equation applied to subsurface drainage by pipes, tile drains or ditches. An alternative subsurface drainage method is drainage by wells for which groundwater flow equations are also available.

Calculation of Groundwater Flow

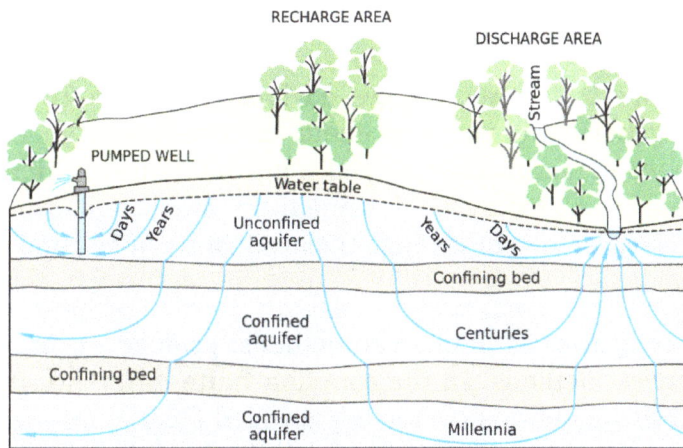

Relative groundwater travel times.

To use the groundwater flow equation to estimate the distribution of hydraulic heads, or the direction and rate of groundwater flow, this partial differential equation (PDE) must be solved. The most common means of analytically solving the diffusion equation in the hydrogeology literature are:

- Laplace, Hankel and Fourier transforms (to reduce the number of dimensions of the PDE),

- similarity transform (also called the Boltzmann transform) is commonly how the Theis solution is derived,

- separation of variables, which is more useful for non-Cartesian coordinates, and

- Green's functions, which is another common method for deriving the Theis solution — from the fundamental solution to the diffusion equation in free space.

No matter which method we use to solve the groundwater flow equation, we need both initial conditions (heads at time (t) = 0) and boundary conditions (representing either the physical boundaries of the domain, or an approximation of the domain beyond that point). Often the initial conditions are supplied to a transient simulation, by a corresponding steady-state simulation (where the time derivative in the groundwater flow equation is set equal to 0).

There are two broad categories of how the (PDE) would be solved; either analytical methods, numerical methods, or something possibly in between. Typically, analytic methods solve the groundwater flow equation under a simplified set of conditions *exactly*, while numerical methods solve it under more general conditions to an *approximation*.

Analytic Methods

Analytic methods typically use the structure of mathematics to arrive at a simple, elegant solution, but the required derivation for all but the simplest domain geometries can be quite complex (involving non-standard coordinates, conformal mapping, etc.). Analytic solutions typically are also simply an equation that can give a quick answer based on a few basic parameters. The Theis equation is a very simple (yet still very useful) analytic solution to the groundwater flow equation, typically used to analyze the results of an aquifer test or slug test.

Numerical Methods

The topic of numerical methods is quite large, obviously being of use to most fields of engineering and science in general. Numerical methods have been around much longer than computers have (In the 1920s Richardson developed some of the finite difference schemes still in use today, but they were calculated by hand, using paper and pencil, by human "calculators"), but they have become very important through the availability of fast and cheap personal computers.

There are two broad categories of numerical methods: gridded or discretized methods and non-gridded or mesh-free methods. In the common finite difference method and finite element method (FEM) the domain is completely gridded ("cut" into a grid or mesh of small elements). The analytic element method (AEM) and the boundary integral equation method (BIEM — sometimes also called BEM, or Boundary Element Method) are only discretized at boundaries or along flow elements (line sinks, area sources, etc.), the majority of the domain is mesh-free.

General Properties of Gridded Methods

Gridded Methods like finite difference and finite element methods solve the groundwater flow equation by breaking the problem area (domain) into many small elements (squares, rectangles, triangles, blocks, tetrahedra, etc.) and solving the flow equation for each element (all material properties are assumed constant or possibly linearly variable within an element), then linking together all the elements using conservation of mass across the boundaries between the elements (similar to the divergence theorem). This results in a system which overall approximates the groundwater flow equation, but exactly matches the boundary conditions (the head or flux is specified in the elements which intersect the boundaries).

Finite differences are a way of representing continuous differential operators using discrete intervals (Δx and Δt), and the finite difference methods are based on these (they are derived from a Taylor series). For example, the first-order time derivative is often approximated using the following forward finite difference, where the subscripts indicate a discrete time location,

$$\frac{\partial h}{\partial t} = h'(t_i) \approx \frac{h_i - h_{i-1}}{\Delta t}$$

The forward finite difference approximation is unconditionally stable, but leads to an implicit set of equations (that must be solved using matrix methods, e.g. LU or Cholesky decomposition). The similar backwards difference is only conditionally stable, but it is explicit and can be used to "march" forward in the time direction, solving one grid node at a time (or possibly in parallel, since one node depends only on its immediate neighbors). Rather than the finite difference method, sometimes the Galerkin FEM approximation is used in space (this is different from the type of FEM often used in structural engineering) with finite differences still used in time.

Application of Finite Difference Models

MODFLOW is a well-known example of a general finite difference groundwater flow model. It is developed by the US Geological Survey as a modular and extensible simulation tool for modeling groundwater flow. It is free software developed, documented and distributed by the USGS. Many commercial products have grown up around it, providing graphical user interfaces to its input file based interface, and typically incorporating pre- and post-processing of user data. Many other models have been developed to work with MODFLOW input and output, making linked models which simulate several hydrologic processes possible (flow and transport models, surface water and groundwater models and chemical reaction models), because of the simple, well documented nature of MODFLOW.

Application of Finite Element Models

Finite Element programs are more flexible in design (triangular elements vs. the block elements most finite difference models use) and there are some programs available (SUTRA, a 2D or 3D density-dependent flow model by the USGS; Hydrus, a commercial unsaturated flow model; FEFLOW, a commercial modelling environment for subsurface flow, solute and heat transport processes; OpenGeoSys, a scientific open-source project for thermo-hydro-mechanical-chemical (THMC) processes in porous and fractured media; COMSOL Multiphysics (FEMLAB) a commercial general modelling environment), FEATool Multiphysics, an easy to use Matlab simulation toolbox, and Integrated Water Flow Model (IWFM), but they are still not as popular in with practicing hydrogeologists as MODFLOW is. Finite element models are more popular in university and laboratory environments, where specialized models solve non-standard forms of the flow equation (unsaturated flow, density dependent flow, coupled heat and groundwater flow, etc.)

Application of Finite Volume Models

The finite volume method is a method for representing and evaluating partial differential equations as algebraic equations. Similar to the finite difference method, values are calculated at discrete places on a meshed geometry. "Finite volume" refers to the small volume surrounding each node point on a mesh. In the finite volume method, volume integrals in a partial differential equation that contain a divergence term are converted to surface integrals, using the divergence theorem. These terms are then evaluated as fluxes at the surfaces of each finite volume. Because the flux entering a given volume is identical to that leaving the adjacent volume, these methods are conservative. Another advantage of the finite volume method is that it is easily formulated to allow for unstructured meshes. The method is used in many computational fluid dynamics packages.

PORFLOW software package is a comprehensive mathematical model for simulation of Ground Water Flow and Nuclear Waste Management developed by Analytic & Computational Research, Inc., ACRi.

The FEHM software package is available free from Los Alamos National Laboratory. This versatile porous flow simulator includes capabilities to model multiphase, thermal, stress, and multicomponent reactive chemistry. Current work using this code includes simulation of methane hydrate formation, CO_2 sequestration, oil shale extraction, migration of both nuclear and chemical contaminants, environmental isotope migration in the unsaturated zone, and karst formation.

Other Methods

These include mesh-free methods like the Analytic Element Method (AEM) and the Boundary Element Method (BEM), which are closer to analytic solutions, but they do approximate the groundwater flow equation in some way. The BEM and AEM exactly solve the groundwater flow equation (perfect mass balance), while approximating the boundary conditions. These methods are more exact and can be much more elegant solutions (like analytic methods are), but have not seen as widespread use outside academic and research groups yet.

Hydroinformatics

Hydroinformatics is a branch of informatics which concentrates on the application of information and communications technologies (ICTs) in addressing the increasingly serious problems of the equitable and efficient use of water for many different purposes. Growing out of the earlier discipline of computational hydraulics, the numerical simulation of water flows and related processes remains a mainstay of hydroinformatics, which encourages a focus not only on the technology but on its application in a social context.

On the technical side, in addition to computational hydraulics, hydroinformatics has a strong interest in the use of techniques originating in the so-called artificial intelligence community, such as artificial neural networks or recently support vector machines and genetic programming. These might be used with large collections of observed data for the purpose of data mining for knowledge discovery, or with data generated from an existing, physically based model in order to generate a computationally efficient emulator of that model for some purpose.

Hydroinformatics recognises the inherently social nature of the problems of water management and of decision making processes, and strives to understand the social processes by which technologies are brought into use. Since the problems of water management are most severe in the majority world, while the resources to obtain and develop technological solutions are concentrated in the hands of the minority, the need to examine these social processes are particularly acute.

Hydroinformatics draws on and integrates hydraulics, hydrology, environmental engineering and many other disciplines. It sees application at all points in the water cycle from atmosphere to ocean, and in artificial interventions in that cycle such as urban drainage and water supply systems. It provides support for decision making at all levels from governance and policy through management to operations.

Hydroinformatics has a growing world-wide community of researchers and practitioners. The Journal of Hydroinformatics provides a specific outlet for Hydroinformatics research, and the community gathers to exchange ideas at the biennial conferences. These activities are coordinated by the joint IAHR, IWA, IAHS Hydroinformatics Section.

There is a growing need for professionals and managers to appreciate and work with these new technologies and tools.

Hydrometeorology

Hydrometeorology is a branch of meteorology and hydrology that studies the transfer of water and energy between the land surface and the lower atmosphere. UNESCO has several programmes and activities in place that deal with the study of natural hazards of hydrometeorological origin and the mitigation of their effects. Among these hazards are the results of natural processes or phenomena of atmospheric, hydrological or oceanographic nature such as floods, tropical cyclones, drought and desertification. Many countries have established an operational hydrometeorological capability to assist with forecasting, warning and informing the public of these developing hazards.

Operational Hydrometeorology in Practice

Countries with a current operational hydrometeorology service include, among others:

- Australia
- Canada
- England and Wales (Flood Forecasting Centre)
- France
- India
- Scotland
- Serbia
- Russia
- The United States of America

Isotope Hydrology

Isotope hydrology is a field of hydrology that uses isotopic dating to estimate the age and origins of water and of movement within the hydrologic cycle. The techniques are used for water-use policy, mapping aquifers, conserving water supplies, and controlling pollution. It replaces or supplements past methods of measuring rain, river levels and other bodies of water over many decades.

Details

Water molecules carry unique fingerprints, based in part on differing proportions of the oxygen and hydrogen isotopes that constitute all water. Isotopes are forms of the same element that have variable numbers of neutrons in their nuclei.

Air, soil and water contain mostly oxygen 16 (^{16}O). Oxygen 18 (^{18}O) occurs in approximately one oxygen atom in every five hundred and is a bit heavier than oxygen 16, as it has two extra neutrons. From a simple energy standpoint this results in a preference for evaporating the lighter ^{16}O containing water and leaving more of the ^{18}O water behind in the liquid state (called fractionation). Thus seawater tends to be richer in ^{18}O and rain and snow relatively depleted in ^{18}O.

Carbon 14 dating is also used as part of isotope hydrology as all natural water contains dissolved carbon dioxide.

Applications

One commonly cited application involves the use of stable isotopes to determine the age of ice or snow, which can help indicate the conditions of the climate in the past. Higher average global temperature would provide more energy and thus an increase in atmospheric ^{18}O water, while lower than normal amounts of ^{18}O in groundwater or an ice layer would imply that the water or ice represents an evaporation origin during cooler climatic eras or even ice ages.

Another application involves the separation of groundwater flow and baseflow from streamflow in the field of catchment hydrology (i.e. a method of hydrograph separation). Since precipitation in each rain or snowfall event has a specific isotopic signature, and the signatures of subsurface water can also be identified by well sampling, the composite signature in the stream is an indicator of, at any given time, what portion of the streamflow comes from overland flow and what portion comes from subsurface flow.

Current Use

The isotope hydrology program at the International Atomic Energy Agency works to aid developing states (including 84 projects in more than 50 countries) and to create a detailed portrait of Earth's water resources.

In Ethiopia, Libya, Chad, Egypt and Sudan, the International Atomic Energy Agency used such techniques to help local water policy deal with fossil water.

An arsenic pollution crisis in Bangladesh that the World Health Organization calls the "largest mass poisoning of a population in history" has been investigated using this technique.

Limnology

Lake Hāwea, New Zealand

Limnology is the study of inland waters. It is often regarded as a division of ecology or environmental science. It covers the biological, chemical, physical, geological, and other attributes of all inland waters (running and standing waters, both fresh and saline, natural or man-made). This includes the study of lakes and ponds, rivers, springs, streams and wetlands. A more recent sub-discipline of limnology, termed landscape limnology, studies, manages, and conserves these aquatic ecosystems using a landscape perspective.

Limnology is closely related to aquatic ecology and hydrobiology, which study aquatic organisms in particular regard to their hydrological environment. Although limnology is sometimes equated with freshwater science, this is erroneous since limnology also comprises the study of inland salt lakes.

History

The term limnology was coined by François-Alphonse Forel (1841–1912) who established the field with his studies of Lake Geneva. Interest in the discipline rapidly expanded, and in 1922 August Thienemann (a German zoologist) and Einar Naumann (a Swedish botanist) co-founded the International Society of Limnology (SIL, from Societas Internationalis Limnologiae). Forel's original definition of limnology, "the oceanography of lakes", was expanded to encompass the study of all inland waters, and influenced Benedykt Dybowski's work on Lake Baikal.

Prominent early American limnologists included G. Evelyn Hutchinson, Ed Deevey, E. A. Birge, and C. Juday.

Lake Classification

Lake George, New York, United States, an oligotrophic lake

Limnology classifies lakes (or other bodies of water) according to the trophic state index. An oligotrophic lake is characterised by relatively low levels of primary production and low levels of nutrients. A eutrophic lake has high levels of primary productivity due to very high nutrient levels. Eutrophication of a lake can lead to algal blooms. Dystrophic lakes have high levels of humic matter and typically has yellow-brown, tea-coloured waters. These categories do not have rigid specifications; the classification system can be seen as more of a spectrum encompassing the various levels of aquatic productivity.

Surface-Water Hydrology

Surface-water hydrology is a field that encompasses all surface waters of the globe (overland flows, rivers, lakes, wetlands, estuaries, oceans, etc.). This is a subset of the hydrologic cycle that does not include atmospheric, and ground waters. Surface-water hydrology relates the dynamics of flow in surface-water systems (rivers, canals, streams, lakes, ponds, wetlands, marshes, arroyos, oceans, etc.). This includes the field measurement of flow (discharge); the statistical variability at each setting; floods; drought susceptibility and the development of the levels of risk; and the fluid mechanics of surface waters.

In-depth analysis of surface-water components of the hydrologic cycle: hydrometeorology, evaporation/transpiration, rainfall-runoff relationships, open-channel flow, flood hydrology, fluid mechanics, and statistical and probabilistic methods in hydrology. Surface-water hydrology includes the relation between rainfall and surface runoff; this relationship is an important aspect of water resources for sewerage (wastewater) or (sewage), drinking water, agriculture (irrigation) environmental protection, and for flood control.

The relationships between groundwater and surface water includes baseflow needs for instream flow, and subsurface water levels in wells.

A stormwater engineer is a civil engineer who manages the flow, filtering, and release of stormwater.

References

- "Guidelines for the Integrated Management of the Watershed – Phytotechnology & Ecohydrology", by Zalewski, M. (2002) (Ed). United Nations Environment Programme Freshwater Management Series No. 5. 188pp, ISBN 92-807-2059-7.

- Ecohydrology of Water-controlled Ecosystems : Soil Moisture and Plant Dynamics, Ignacio Rodríguez-Iturbe, Amilcare Porporato, 2005. ISBN 0-521-81943-1

- Good, Stephen P.; Noone, David; Bowen, Gabriel (2015-07-10). "Hydrologic connectivity constrains partitioning of global terrestrial water fluxes". Science. 349 (6244): 175–177. doi:10.1126/science.aaa5931. ISSN 0036-8075.

- "Hydro-meteorological hazards | United Nations Educational, Scientific and Cultural Organization". Unesco. org. Retrieved 2014-05-28.

- "Republic Hydrometeorological service of Serbia Kneza Višeslava 66 Beograd". Hidmet.gov.rs. 2014-05-18. Retrieved 2014-05-28.

Essential Aspects of Hydrology

Fundamental features analyzed by hydrology include infiltration, routing, evapotranspiration and water quality. This chapter focuses on the characteristics, calculating methods and terminology of each feature. A section of the text discusses hydrograph and hydrography. This chapter provides a plethora of topics for better comprehension of hydrology.

Infiltration (Hydrology)

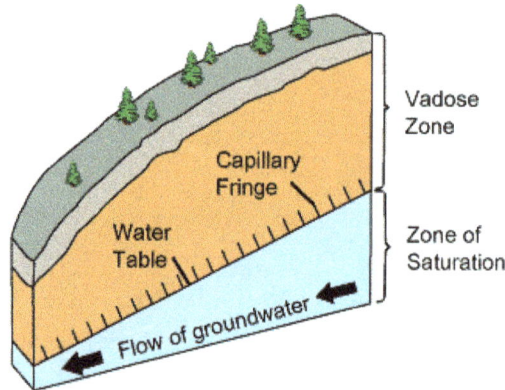

Cross-section of a hillslope depicting the vadose zone, capillary fringe, water table, and phreatic or saturated zone. (Source: United States Geological Survey.)

Infiltration is the process by which water on the ground surface enters the soil. *Infiltration rate* in soil science is a measure of the rate at which soil is able to absorb rainfall or irrigation. It is measured in inches per hour or millimeters per hour. The rate decreases as the soil becomes saturated. If the precipitation rate exceeds the infiltration rate, runoff will usually occur unless there is some physical barrier. It is related to the saturated hydraulic conductivity of the near-surface soil. The rate of infiltration can be measured using an infiltrometer.

Introduction

Infiltration is caused by two forces: gravity and capillary action. While smaller pores offer greater resistance to gravity, very small pores pull water through capillary action in addition to and even against the force of gravity.

The rate of infiltration is determined by soil characteristics including ease of entry, storage capacity, and transmission rate through the soil. The soil texture and structure, vegetation types and cover, water content of the soil, soil temperature, and rainfall intensity all play a role in controlling infiltration rate and capacity. For example, coarse-grained sandy soils have large spaces between each grain and allow water to infiltrate quickly. Vegetation creates more porous soils by both protecting the soil from raindrop impact, which can close natural gaps between soil particles, and loosening soil through root action. This is why forested areas have the highest infiltration rates of any vegetative types.

The top layer of leaf litter that is not decomposed protects the soil from the pounding action of rain; without this the soil can become far less permeable. In chaparral vegetated areas, the hydrophobic oils in the succulent leaves can be spread over the soil surface with fire, creating large areas of hydrophobic soil. Other conditions that can lower infiltration rates or block them include dry plant litter that resists re-wetting, or frost. If soil is saturated at the time of an intense freezing period, the soil can become a concrete frost on which almost no infiltration would occur. Over an entire watershed, there are likely to be gaps in the concrete frost or hygroscopic soil where water can infiltrate.

Once water has infiltrated the soil it remains in the soil, percolates down to the ground water table, or becomes part of the subsurface runoff process.

Process

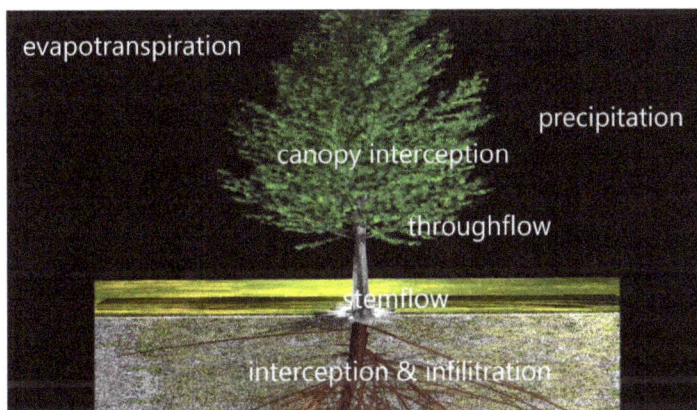

The process of infiltration can continue only if there is room available for additional water at the soil surface. The available volume for additional water in the soil depends on the porosity of the soil and the rate at which previously infiltrated water can move away from the surface through the soil. The maximum rate that water can enter a soil in a given condition is the infiltration capacity. If the arrival of the water at the soil surface is less than the infiltration capacity, it is sometimes analyzed using hydrology transport models, mathematical models that consider infiltration, runoff and channel flow to predict river flow rates and stream water quality.

Research Findings

Robert E. Horton suggested that infiltration capacity rapidly declines during the early part of a storm and then tends towards an approximately constant value after a couple of hours for the remainder of the event. Previously infiltrated water fills the available storage spaces and reduces the capillary forces drawing water into the pores. Clay particles in the soil may swell as they become wet and thereby reduce the size of the pores. In areas where the ground is not protected by a layer of forest litter, raindrops can detach soil particles from the surface and wash fine particles into surface pores where they can impede the infiltration process.

Infiltration in Wastewater Collection

Wastewater collection systems consist of a set of lines, junctions and lift stations to convey sewage to a wastewater treatment plant. When these Herr lines are compromised by rupture, cracking or tree root invasion, infiltration/inflow of stormwater often occurs. This circumstance can lead to a sanitary sewer overflow, or discharge of untreated sewage to the environment.

Infiltration Calculation Methods

Infiltration is a component of the general mass balance hydrologic budget. There are several ways to estimate the volume and/or the rate of infiltration of water into a soil. The rigorous standard that fully couples groundwater to surface water through a non-homogeneous soil is the numerical solution of Richards' equation. A newer method that allows full groundwater and surface water coupling in homogeneous soil layers, and that is related to the Richards equation is the Finite water-content vadose zone flow method. In the case of uniform initial soil water content and a deep well-drained soil, there are some excellent approximate methods to solve for the infiltration flux for a single rainfall event. Among these are the Green and Ampt (1911) method, Parlange et al. (1982). Beyond these methods there are a host of empirical methods such as, SCS method, Horton's method, etc., that are little more than curve fitting exercises.

General Hydrologic Budget

The general hydrologic budget, with all the components, with respect to infiltration F. Given all the other variables and infiltration is the only unknown, simple algebra solves the infiltration question.

$$F = B_I + P - E - T - ET - S - I_A - R - B_O$$

where

F is infiltration, which can be measured as a volume or length;

B_I is the boundary input, which is essentially the output watershed from adjacent, directly connected impervious areas;

B_O is the boundary output, which is also related to surface runoff, R, depending on where one chooses to define the exit point or points for the boundary output;

P is precipitation;

E is evaporation;

T is transpiration;

ET is evapotranspiration;

S is the storage through either retention or detention areas;

I_A is the initial abstraction, which is the short term surface storage such as puddles or even possibly detention ponds depending on size;

R is surface runoff.

The only note on this method is one must be wise about which variables to use and which to omit, for doubles can easily be encountered. An easy example of double counting variables is when the evaporation, *E*, and the transpiration, *T*, are placed in the equation as well as the evapotranspiration, *ET*. *ET* has included in it *T* as well as a portion of *E*. Interception also needs to be accounted for, not just raw precipitation.

Richards' Equation (1931)

The standard rigorous approach for calculating infiltration into soils is Richards' Equation, which is a partial differential equation with very nonlinear coefficients. The Richards equation is computationally expensive, not guaranteed to converge, and sometimes has difficulty with mass conservation.

Finite Water-Content Vadose Zone Flow Method

This method is an approximation of the Richards' (1931) partial differential equation that de-emphasized soil water diffusivity and emphasizes advection. This approximation does not affect the calculated infiltration flux because the diffusive flux has a mean of 0. The Finite water-content vadose zone flow method is a set of three ordinary differential equations, is guaranteed to converge and to conserve mass. It requires the assumption that soil is uniform within layers.

Green and Ampt

Named for two men; Green and Ampt. The Green-Ampt method of infiltration estimation accounts for many variables that other methods, such as Darcy's law, do not. It is a function of the soil suction head, porosity, hydraulic conductivity and time.

$$\int_0^{F(t)} \frac{F}{F + \psi \Delta \theta} dF = \int_0^t K \, dt$$

where

ψ is wetting front soil suction head (L);

θ is water content (-);

K is Hydraulic conductivity (L/T);

$F(t)$ is the cumulative depth of infiltration (L).

Once integrated, one can easily choose to solve for either volume of infiltration or instantaneous infiltration rate:

$$F(t) = Kt + \psi \Delta \theta \ln \left[1 + \frac{F(t)}{\psi \Delta \theta} \right]$$

Using this model one can find the volume easily by solving for $F(t)$. However the variable being solved for is in the equation itself so when solving for this one must set the variable in question to converge on zero, or another appropriate constant. A good first guess for F is the larger value of either Kt or $\sqrt{2\psi \Delta\theta K t}$. The only note on using this formula is that one must assume that h_0 , the water head or the depth of ponded water above the surface, is negligible. Using the infiltration volume from this equation one may then substitute F into the corresponding infiltration rate equation below to find the instantaneous infiltration rate at the time, t, F was measured.

$$f(t) = K\left[\frac{\psi \Delta\theta}{F(t)} + 1\right].$$

Horton's Equation

Named after the same Robert E. Horton mentioned above, Horton's equation is another viable option when measuring ground infiltration rates or volumes. It is an empirical formula that says that infiltration starts at a constant rate, f_0, and is decreasing exponentially with time, t. After some time when the soil saturation level reaches a certain value, the rate of infiltration will level off to the rate f_c.

$$f_t = f_c + (f_0 - f_c)e^{-kt}$$

Where

f_t is the infiltration rate at time t;

f_0 is the initial infiltration rate or maximum infiltration rate;

f_c is the constant or equilibrium infiltration rate after the soil has been saturated or minimum infiltration rate;

k is the decay constant specific to the soil.

The other method of using Horton's equation is as below. It can be used to find the total volume of infiltration, F, after time t.

$$F_t = f_c t + \frac{(f_0 - f_c)}{k}(1 - e^{-kt})$$

Kostiakov Equation

Named after its founder Kostiakov is an empirical equation which assumes that the intake rate declines over time according to a power function.

$$f(t) = akt^{a-1}$$

Where a and k are empirical parameters.

The major limitation of this expression is its reliance on the zero final intake rate. In most cases the infiltration rate instead approaches a finite steady value, which in some cases may occur after short periods of time. The Kostiakov-Lewis variant, also known as the "Modified Kostiakov" equation corrects for this by adding a steady intake term to the original equation.

$$f(t) = akt^{a-1} + f_0$$

in integrated form the cumulative volume is expressed as:

$$F(t) = kt^a + f_0 t$$

Where

f_0 approximates, but does not necessarily equate to the final infiltration rate of the soil.

Darcy's Law

This method used for infiltration is using a simplified version of Darcy's law. Many would argue that this method is too simple and should not be used. Compare it with the Green and Ampt (1911) solution mentioned previously. This method is similar to Green and Ampt, but missing the cumulative infiltration depth and is therefore incomplete because it assumes that the infiltration gradient occurs over some arbitrary length f_0. In this model the ponded water is assumed to be equal to h_0 and the head of dry soil that exists below the depth of the wetting front soil suction head is assumed to be equal to $-\psi - L$.

$$f = K\left[\frac{h_0 - (-\psi - L)}{L}\right]$$

where

ψ is wetting front soil suction head

h_0 is the depth of ponded water above the ground surface;

K is the hydraulic conductivity;

L is the vague total depth of subsurface ground in question. This vague definition explains why this method should be avoided.

or

$$f = K\left[\frac{L + S_f + h_0}{L}\right]$$

f Infiltration rate f (mm hour$^{-1)}$)

K is the hydraulic conductivity (mm hour$^{-1)}$);

L is the vague total depth of subsurface ground in question (mm). This vague definition explains why this method should be avoided.

S_f is wetting front soil suction head $(-\psi) = (-\psi_f)$ (mm)

h_0 is the depth of ponded water above the ground surface (mm);

Routing (Hydrology)

In hydrology, routing is a technique used to predict the changes in shape of a hydrograph as water moves through a river channel or a reservoir. In flood forecasting, hydrologists may want to know how a short burst of intense rain in an area upstream of a city will change as it reaches the city. Routing can be used to determine whether the pulse of rain reaches the city as a deluge or a trickle.

Routing also can be used to predict the hydrograph shape (and thus lowland flooding potential) subsequent to multiple rainfall events in different sub-catchments of the watershed. Timing and duration of the rainfall events, as well as factors such as antecedent moisture conditions, overall watershed shape, along with subcatchment-area shapes, land slopes (topography/physiography), geology/hydrogcology (i.e. forests and aquifers can serve as giant sponges that absorb rainfall and slowly release it over subsequent weeks and months), and stream-reach lengths all play a role here. The result can be an additive effect (i.e. a large flood if each subcatchment's respective hydrograph peak arrives at the watershed mouth at the same point in time, thereby effectively causing a "stacking" of the hydrograph peaks), or a more distributed-in-time effect (i.e. a lengthy but relatively modest flood, effectively attenuated in time, as the individual subcatchment peaks arrive at the mouth of the main watershed channel in orderly succession).

Other uses of routing include reservoir and channel design, floodplain studies and watershed simulations.

If the water flow at a particular point, A, in a stream is measured over time with a flow gauge, this information can be used to create a hydrograph. A short period of intense rain, normally called a flood event, can cause a bulge in the graph, as the increased water travels down the river, reaches the flow gauge at A, and passes along it.

If another flow gauge at B, downstream of A is set up, one would expect the graph's bulge (or floodwave) to have the same shape. However, the shape of the river and flow resistance within a river (from the river bed, for example) can affect the shape of the floodwave. Oftentimes, the floodwave will be attenuated (have a reduced peak flow).

Routing techniques can be broadly classified as follows:

Hydraulic (or Distributed) Routing

Hydraulic routing is based on the solution of partial differential equations of unsteady open-channel flow. The equations used are the St. Venant equations or the dynamic wave equations.

The hydraulic models (e.g. *dynamic* and *diffusion* wave models) require the gathering of a lot of data related to river geometry and morphology and consume a lot of computer resources in order to solve the Saint-Venant equations numerically.

Hydrologic (or Lumped) Routing

Hydrologic routing uses the continuity equation for hydrology. In its simplest form, inflow to the river reach is equal to the outflow of the river reach plus the change of storage:

$$I = O + \frac{\Delta S}{\Delta t}, \text{ where}$$

- *I* is average inflow to the reach during Ät

- *O* is average outflow from the reach during Ät ; and

- *S* is the water currently in the reach (known as storage)

The hydrologic models (e.g. *linear* and *nonlinear Muskingum* models) need to estimate hydrologic parameters using recorded data in both upstream and downstream sections of rivers and/or by applying robust optimization techniques to solve the one-dimensional conservation of mass and storage-continuity equation.

Semi-distributed Routing

Nowadays, semi-distributed models such as Muskingum–Cunge family procedures are also available. Simple physical concepts and common river characteristics such as channel geometry, reach length, roughness coefficient, and slope are used to estimate the model parameters without complex and expensive numerical solutions.

Selection of Routing Procedure

In general, based on the available field data and goals of the project, one of routing procedures is selected.

Hydrograph

A hydrograph is a graph showing the rate of flow (discharge) versus time past a specific point in a river, or other channel or conduit carrying flow. The rate of flow is typically expressed in cubic meters or cubic feet per second (cms or cfs).

It can also refer to a graph showing the volume of water reaching a particular outfall, or location in a sewerage network. Graphs are commonly used in the design of sewerage, more specifically, the design of surface water sewerage systems and combined sewers.

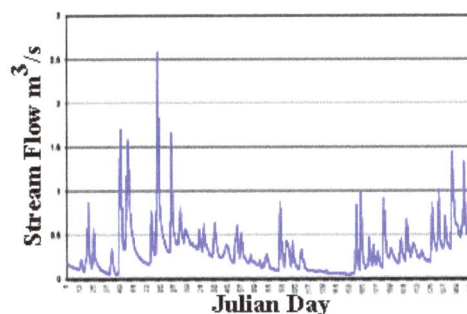

Stream hydrograph. Increases in stream flow follow rainfall or snowmelt. The gradual decay in flow after the peaks reflects diminishing supply from groundwater.

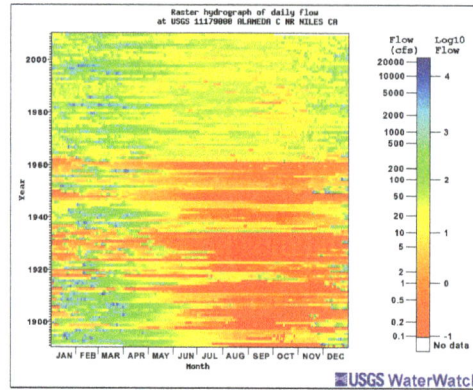

Raster hydrograph. The entire flow record and patterns representing different timescales can be visualized.

Terminology

The discharge is measured at a specific point in a river and is typically time variant.

- Rising limb: The rising limb of hydro graph, also known as concentration curve, reflects a prolonged increase in discharge from a catchment area, typically in response to a rainfall event

- Recession (or falling) limb: The recession limb extends from the peak flow rate onward. The end of stormflow (aka quickflow or direct runoff) and the return to groundwater-derived flow (base flow) is often taken as the point of inflection of the recession limb. The recession limb represents the withdrawal of water from the storage built up in the basin during the earlier phases of the hydrograph.

- Peak discharge: the highest point on the hydro graph when the rate of discharge is greatest

- Lag time: the time interval from the center of mass of rainfall excess to the peak of the resulting hydrograph

- Time to peak: time interval from the start of the resulting hydro graph

- Discharge: the rate of flow (volume per unit time) passing a specific location in a river or other channel

Types of hydrograph can include:

- Storm hydrographs

- Flood hydrographs

- Annual hydrographs aka regimes

- Direct Runoff Hydrograph

- Effective Runoff Hydrograph

- Raster Hydrograph

- Storage opportunities in the drainage network (e.g., lakes, reservoirs, wetlands, channel and bank storage capacity)

Baseflow Separation

A stream hydrograph is commonly conceptualized to include a baseflow component' *and a* runoff component. The former represents the relatively steady contribution to stream discharge from groundwater return flow, while the latter represents the additional streamflow contributed by sub-surface flow and surface flow/runoff.

The separation of baseflow from direct runoff in a hydrograph is often of interest to hydrologists, planners, and engineers, as it aids in determining the influence of different hydrologic processes on discharge from the subject catchment. Because the timing, magnitude, and duration of groundwater return flow differs so greatly from that of direct runoff, separating and understanding the influence of these distinct processes is key to analyzing and simulating the likely hydrologic effects of various land use, water use, weather, and climate conditions and changes.

However, the process of separating "baseflow" from "direct runoff" is an inexact science. In part this is because these two concepts are not, themselves, entirely distinct and unrelated. Return flow from groundwater increases along with overland flow from saturated or impermeable areas during and after a storm event; moreover, a particular water molecule can easily move through both pathways en route to the watershed outlet. Therefore, separation of a purely "baseflow component" in a hydrograph is a somewhat arbitrary exercise. Nevertheless, various graphical and empirical techniques have been developed to perform these hydrograph separations. The separation of base flow from direct runoff can be an important first step in developing rainfall-runoff models for a watershed of interest—for example, in developing and applying unit hydrographs as described below.

Unit Hydrograph

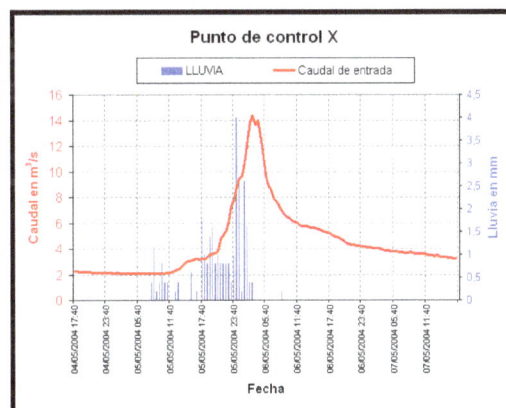

A *unit hydrograph* (UH) is the hypothetical unit response of a watershed (in terms of runoff volume and timing) to a unit input of rainfall. It can be defined as the *direct runoff hydrograph* (DRH) resulting from one unit (e.g., one cm or one inch) of *effective rainfall* occurring uniformly over that watershed at a uniform rate over a unit period of time. As a UH is applicable only to the direct runoff component of a hydrograph (i.e., surface runoff), a separate determination of the baseflow component is required.

A UH is specific to a particular watershed, and specific to a particular length of time corresponding to the duration of the effective rainfall. That is, the UH is specified as being the 1-hour, 6-hour, or 24-hour UH, or any other length of time up to the *time of concentration* of direct runoff at the watershed outlet. Thus, for a given watershed, there can be many unit hydrographs, each one corresponding to a different duration of effective rainfall.

The UH technique provides a practical and relatively easy-to-apply tool for quantifying the effect of a unit of rainfall on the corresponding runoff from a particular drainage basin. UH theory assumes that a watershed's runoff response is linear and time-invariant, and that the effective rainfall occurs uniformly over the watershed. In the real world, none of these assumptions are strictly true. Nevertheless, application of UH methods typically yields a reasonable approximation of the flood response of natural watersheds. The linear assumptions underlying UH theory allows for the variation in storm intensity over time (i.e., the storm *hyetograph*) to be simulated by applying the principles of superposition and proportionality to separate storm components to determine the resulting cumulative hydrograph. This allows for a relatively straightforward calculation of the hydrograph response to any arbitrary rain event.

An instantaneous unit hydrograph is a further refinement of the concept; for an IUH, the input rainfall is assumed to all take place at a discrete point in time (obviously, this isn't the case for actual rainstorms). Making this assumption can greatly simplify the analysis involved in constructing a unit hydrograph, and it is necessary for the creation of a geomorphologic instantaneous unit hydrograph.

The creation of a GIUH is possible given nothing more than topologic data for a particular drainage basin. In fact, only the number of streams of a given order, the mean length of streams of a given order, and the mean land area draining directly to streams of a given order are absolutely required (and can be estimated rather than explicitly calculated if necessary). It is therefore possible to calculate a GIUH for a basin without any data about stream height or flow, which may not always be available.

Subsurface Hydrology Hydrograph

In subsurface hydrology (hydrogeology), a hydrograph is a record of the water level (the observed hydraulic head in wells screened across an aquifer).

Typically, a hydrograph is recorded for monitoring of heads in aquifers during non-test conditions (e.g., to observe the seasonal fluctuations in an aquifer). When an aquifer test is being performed, the resulting observations are typically called drawdown, since they are subtracted from pre-test levels and often only the change in water level is dealt with.

Raster Hydrograph

Raster hydrographs are pixel-based plots for visualizing and identifying variations and changes in large multidimensional data sets. Originally developed by Keim (2000) they were first applied in hydrology by Koehler (2004) as a means of highlighting inter-annual and intra-annual changes in streamflow. The raster hydrographs in WaterWatch, like those developed by Koehler, depict years on the y-axis and days along the x-axis. Users can choose to plot streamflow (actual values or log values), streamflow percentile, or streamflow class (from 1, for low flow, to 7 for high flow), for Daily, 7-Day, 14-Day, and 28-Day streamflow. For a more comprehensive description of raster hydrographs.

Hydrography

HMS Waterwitch, a hydrographic survey vessel

Hydrography is the branch of applied sciences which deals with the measurement and description of the physical features of oceans, seas, coastal areas, lakes and rivers, as well as with the prediction of their change over time, for the primary purpose of safety of navigation and in support of all other marine activities, including economic development, security and defence, scientific research, and environmental protection.

History

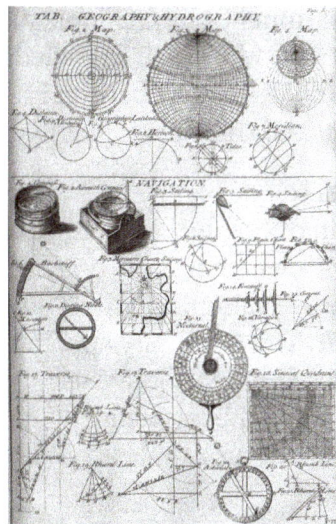

Table of geography, hydrography, and navigation, from the 1728 *Cyclopaedia*.

Survey of the strategic port of Milford Haven produced by Lewis Morris in 1748

Alexander Dalrymple, the first Hydrographer of the Navy in the United Kingdom, appointed in 1795.

The origins of hydrography lay in the making of charts to aid navigation, by individual mariners as they navigated into new waters. These were usually the private property, even closely held secrets, of individuals who used them for commercial or military advantage. As transoceanic trade and exploration increased, hydrographic surveys started to be carried out as an exercise in their own right, and the commissioning of surveys was increasingly done by governments and special hydrographic offices. National organizations, particularly navies, realized that the collection, systematization and distribution of this knowledge gave it great organizational and military advantages. Thus were born dedicated national hydrographic organizations for the collection, organization, publication and distribution of hydrography incorporated into charts and sailing directions.

Prior to the establishment of the United Kingdom Hydrographic Office, Royal Navy captains were responsible for the provision of their own charts. In practice this meant that ships often sailed with inadequate information for safe navigation, and that when new areas were surveyed, the data rarely reached all those who needed it. The Admiralty appointed Alexander Dalrymple as Hydrographer in 1795, with a remit to gather and distribute charts to HM Ships. Within a year existing charts from the previous two centuries had been collated, and the first catalogue published. The first chart produced under the direction of the Admiralty, was a chart of Quiberon Bay in Brittany, and it appeared in 1800.

Under Captain Thomas Hurd the department received its first professional guidelines and the first catalogues were published and made available to the public and to other nations as well. In 1829, Rear-Admiral Sir Francis Beaufort, as Hydrographer, developed the eponymous Scale, and introduced the first official tide tables in 1833 and the first "Notices to Mariners" in 1834. The Hydrographic Office underwent steady expansion throughout the 19th century; by 1855, the Chart Catalogue listed 1,981 charts giving a definitive coverage over the entire world, and produced over 130,000 charts annually, of which about half were sold.

The word *hydrography* comes from the Ancient Greek δωρ (*hydor*), "water" and γράφω (*graphō*), "to write".

Overview

Large-scale hydrography is usually undertaken by national or international organizations which sponsor data collection through precise surveys and publish charts and descriptive material for navigational purposes. The science of oceanography is, in part, an outgrowth of classical hydrography. In many respects the data are interchangeable, but marine hydrographic data will be particularly directed toward marine navigation and safety of that navigation. Marine resource exploration and exploitation is a significant application of hydrography, principally focused on the search for hydrocarbons.

Hydrographical measurements include the tidal, current and wave information of physical oceanography. They include bottom measurements, with particular emphasis on those marine geographical features that pose a hazard to navigation such as rocks, shoals, reefs and other features that obstruct ship passage. Bottom measurements also include collection of the nature of the bottom as it pertains to effective anchoring. Unlike oceanography, hydrography will include shore features, natural and manmade, that aid in navigation. Therefore, a hydrographic survey may include the accurate positions and representations of hills, mountains and even lights and towers that will aid in fixing a ship's position, as well as the physical aspects of the sea and seabed.

Hydrography, mostly for reasons of safety, adopted a number of conventions that have affected its portrayal of the data on nautical charts. For example, hydrographic charts are designed to portray what is safe for navigation, and therefore will usually tend to maintain least depths and occasionally de-emphasize the actual submarine topography that would be portrayed on bathymetric charts. The former are the mariner's tools to avoid accident. The latter are best representations of the actual seabed, as in a topographic map, for scientific and other purposes. Trends in hydrographic practice since c. 2003–2005 have led to a narrowing of this difference, with many more hydrographic offices maintaining "best observed" databases, and then making navigationally "safe" products as required. This has been coupled with a preference for multi-use surveys, so that the same data collected for nautical charting purposes can also be used for bathymetric portrayal.

Even though, in places, hydrographic survey data may be collected in sufficient detail to portray bottom topography in some areas, hydrographic charts only show depth information relevant for safe navigation and should not be considered as a product that accurately portrays the actual shape of the bottom. The soundings selected from the raw source depth data for placement on the nautical chart are selected for safe navigation and are biased to show predominately the shallowest depths that relate to safe navigation. For instance, if there is a deep area that can not be reached because it is surrounded by shallow water, the deep area may not be shown. The color filled areas that show different ranges of shallow water are not the equivalent of contours on a topographic map since they are often drawn seaward of the actual shallowest depth portrayed. A bathymetric chart does show marine topology accurately. Details covering the above limitations can be found in Part 1 of Bowditch's American Practical Navigator. Another concept that affects safe navigation is the sparsity of detailed depth data from high resolution sonar systems. In more remote areas, the only available depth information has been collected with lead lines. This collection method drops a weighted line to the bottom at intervals and records the depth, often from a rowboat or sail boat. There is no data between soundings or between sounding lines to guarantee that there is not a hazard such as a wreck or a coral head waiting there to ruin a sailor's day. Often, the navigation of the collecting boat does not match today's GPS navigational accuracies. The hydrographic chart will use the best data available and will caveat its nature in a caution note or in the legend of the chart.

A hydrographic survey is quite different from a bathymetric survey in some important respects, particularly in a bias toward least depths due to the safety requirements of the former and geomorphologic descriptive requirements of the latter. Historically, this could include echosoundings being conducted under settings biased toward least depths, but in modern practice hydrographic surveys typically attempt to best measure the depths observed, with the adjustments for navigational safety being applied after the fact.

Hydrography of streams will include information on the stream bed, flows, water quality and surrounding land. Basin or interior hydrography pays special attention to rivers and potable water although if collected data is not for ship navigational uses, and is intended for scientific usage, it is more commonly called *hydrology*.

Hydrography of rivers and streams is also an integral part of water management. Most reservoirs in the United States use dedicated stream gauging and rating tables to determine inflows into the reservoir and outflows to irrigation districts, water municipalities and other users of captured water. River/stream hydrographers use handheld and bank mounted devices, to capture a sectional flow rate of moving water through a section.

Organizations

Hydrographic services in most countries are carried out by specialised hydrographic offices. The international coordination of hydrographic efforts lies with the International Hydrographic Organization.

The United Kingdom Hydrographic Office is one of the oldest, supplying a wide range of charts covering the globe to other countries, allied military organisations and the public.

In the United States, the hydrographic charting function has been carried out since 1807 by the Office of Coast Survey of the National Oceanic and Atmospheric Administration within the U.S. Department of Commerce.

Evapotranspiration

Evapotranspiration (ET) is the sum of evaporation and plant transpiration from the Earth's land and ocean surface to the atmosphere. Evaporation accounts for the movement of water to the air from sources such as the soil, canopy interception, and waterbodies. Transpiration accounts for the movement of water within a plant and the subsequent loss of water as vapor through stomata in its leaves. Evapotranspiration is an important part of the water cycle. An element (such as a tree) that contributes to evapotranspiration can be called an evapotranspirator.

Reference evapotranspiration (ET_0), sometimes incorrectly referred to as potential ET, is a representation of the environmental demand for evapotranspiration and represents the evapotranspiration rate of a short green crop (grass), completely shading the ground, of uniform height and with adequate water status in the soil profile. It is a reflection of the energy available to evaporate water, and of the wind available to transport the water vapour from the ground up into the lower atmosphere. Actual evapotranspiration is said to equal reference evapotranspiration when there is ample water. Some US states utilize a full cover alfalfa reference crop that is 0.5 m in height, rather than the short green grass reference, due to the higher value of ET from the alfalfa reference.

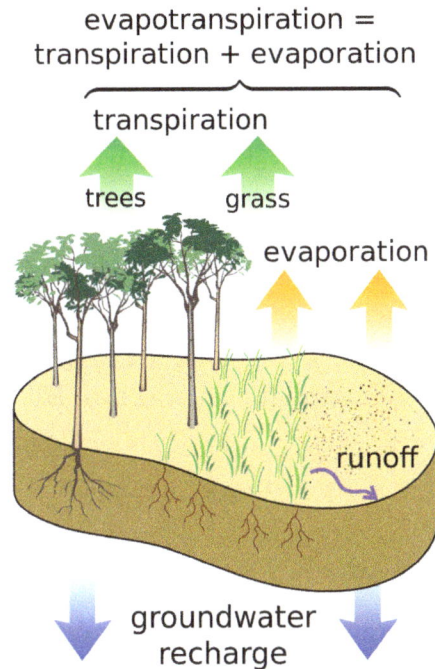

Water cycle of the Earth's surface, showing the individual components of transpiration and evaporation that make up evapotranspiration. Other closely related processes shown are runoff and groundwater recharge.

Water Cycle

Through evapotranspiration, forests reduce water yield, except in unique ecosystems called cloud forests. Trees in cloud forests collect the liquid water in fog or low clouds onto their surface, which drips down to the ground. These trees still contribute to evapotranspiration, but often collect more water than they evaporate or transpire.

In areas that are not irrigated, actual evapotranspiration is usually no greater than precipitation, with some buffer in time depending on the soil's ability to hold water. It will usually be less because some water will be lost due to percolation or surface runoff. An exception is areas with high water tables, where capillary action can cause water from the groundwater to rise through the soil matrix to the surface. If potential evapotranspiration is greater than actual precipitation, then soil will dry out, unless irrigation is used.

Evapotranspiration can never be greater than PET, but can be lower if there is not enough water to be evaporated or plants are unable to transpire readily.

Estimating Evapotranspiration

Evapotranspiration can be measured or estimated using several methods.

Indirect Methods

Pan evaporation data can be used to estimate lake evaporation, but transpiration and evaporation of intercepted rain on vegetation are unknown. There are three general approaches to estimate evapotranspiration indirectly.

Catchment Water Balance

Evapotranspiration may be estimated by creating an equation of the water balance of a drainage basin. The equation balances the change in water stored within the basin (S) with inputs and outgoes:

$$\Delta S = P - ET - Q - D$$

The input is precipitation (P) and the outgoes are evapotranspiration (which is to be estimated), streamflow (Q), and groundwater recharge (D). If the change in storage, precipitation, streamflow, and groundwater recharge are all estimated, the missing flux, ET, can be estimated by rearranging the above equation as follows:

$$ET = P - \Delta S - Q - D$$

Hydrometeorological Equations

The most general and widely used equation for calculating reference ET is the Penman equation. The Penman-Monteith variation is recommended by the Food and Agriculture Organization and the American Society of Civil Engineers. The simpler Blaney-Criddle equation was popular in the Western United States for many years but it is not as accurate in regions with higher humidities. Other solutions used includes Makkink, which is simple but must be calibrated to a specific location, and Hargreaves. To convert the reference evapotranspiration to actual crop evapotranspiration, a crop coefficient and a stress coefficient must be used. Crop coefficients referred to in many hydrological models are themselves during periods for which the model is used. This is because crops are seasonal, perennial plants mature over multiple seasons, and stress responses can significantly depend upon many aspects of plant condition.

Energy Balance

A third methodology to estimate the actual evapotranspiration is the use of the energy balance.

$$\lambda E = R_n - G - H$$

where λE is the energy needed to change the phase of water from liquid to gas, R_n is the net radiation, G is the soil heat flux and H is the sensible heat flux. Using instruments like a scintillometer, soil heat flux plates or radiation meters, the components of the energy balance can be calculated and the energy available for actual evapotranspiration can be solved.

The SEBAL and METRIC algorithms solve the energy balance at the earth's surface using satellite imagery. This allows for both actual and potential evapotranspiration to be calculated on a pixel-by-pixel basis. Evapotranspiration is a key indicator for water management and irrigation performance. SEBAL and METRIC can map these key indicators in time and space, for days, weeks or years.

Experimental Methods for Measuring Evapotranspiration

One method for measuring evapotranspiration is with a weighing lysimeter. The weight of a soil column is measured continuously and the change in storage of water in the soil is modeled by the change in weight. The change in weight is converted to units of length using the surface area of the weighing lysimeter and the unit weight of water. evapotranspiration is computed as the change in weight plus rainfall minus percolation.

Remote Sensing

In recent decades, estimating evapotranspiration has been improved by advances in remote sensing, particularly in agricultural studies. However, quantifying evapotranspiration from mixed vegetation environs, particularly urban parklands, is still challenging because of the heterogeneity of plant species, canopy covers and microclimates and because the methodology is costly. Different remote sensing-based approaches for estimating evapotranspiration have various advantages and disadvantages.

Eddy Covariance

The most direct method of measuring evapotranspiration is with the eddy covariance technique in which fast fluctuations of vertical wind speed are correlated with fast fluctuations in atmospheric water vapor density. This directly estimates the transfer of water vapor (evapotranspiration) from the land (or canopy) surface to the atmosphere.

Urban Landscape Plants

Methods for measuring evapotranspiration can be adapted to an urban setting to estimate the water requirements of urban landscape vegetation.

Potential Evapotranspiration

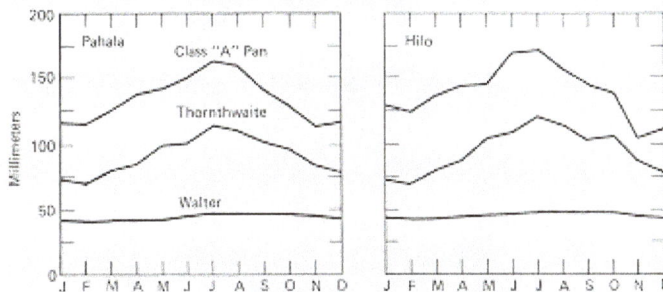

Monthly estimated potential evapotranspiration and measured pan evaporation for two locations in Hawaii, Hilo and Pahala.

Potential evapotranspiration (PET) is the amount of water that would be evaporated and transpired if there were sufficient water available. This demand incorporates the energy available for evaporation and the ability of the lower atmosphere to transport evaporated moisture away from the land surface. Potential evapotranspiration is higher in the summer, on less cloudy days, and closer to the equator, because of the higher levels of solar radiation that provides the energy for evaporation. Potential evapotranspiration is also higher on windy days because the evaporated moisture can be quickly moved from the ground or plant surface, allowing more evaporation to fill its place.

Potential evapotranspiration is expressed in terms of a depth of water, and can be graphed during the year.

Potential evapotranspiration is usually measured indirectly, from other climatic factors, but also depends on the surface type, such as free water (for lakes and oceans), the soil type for bare soil, and the vegetation. Often a value for the potential evapotranspiration is calculated at a nearby climate station on a reference surface, conventionally short grass. This value is called the reference evapotranspiration, and can be converted to a potential evapotranspiration by multiplying with a surface coefficient. In agriculture, this is called a crop coefficient. The difference between potential evapotranspiration and precipitation is used in irrigation scheduling.

Average annual potential evapotranspiration is often compared to average annual precipitation, P. The ratio of the two, P/PET, is the aridity index.

Water Quality

A rosette sampler is used to collect water samples in deep water, such as the Great Lakes or oceans, for water quality testing.

Water quality refers to the chemical, physical, biological, and radiological characteristics of water. It is a measure of the condition of water relative to the requirements of one or more biotic species and or to any human need or purpose. It is most frequently used by reference to a set of standards against which compliance can be assessed. The most common standards used to assess water quality relate to health of ecosystems, safety of human contact, and drinking water.

Standards

In the setting of standards, agencies make political and technical/scientific decisions about how the water will be used. In the case of natural water bodies, they also make some reasonable estimate of pristine conditions. Different uses raise different concerns and therefore different standards are considered. Natural water bodies will vary in response to environmental conditions. Environmental scientists work to understand how these systems function, which in turn helps to identify the sources and fates of contaminants. Environmental lawyers and policymakers work to define legislation with the intention that water is maintained at an appropriate quality for its identified use.

The vast majority of surface water on the planet is neither potable nor toxic. This remains true when seawater in the oceans (which is too salty to drink) is not counted. Another general perception of *water quality* is that of a simple property that tells whether water is polluted or not. In fact, water quality is a complex subject, in part because water is a complex medium intrinsically tied to the ecology of the Earth. Industrial and commercial activities (e.g. manufacturing, mining, construction, transport) are a major cause of water pollution as are runoff from agricultural areas, urban runoff and discharge of treated and untreated sewage.

Categories

The parameters for water quality are determined by the intended use. Work in the area of water quality tends to be focused on water that is treated for human consumption, industrial use, or in the environment.

Human Consumption

Contaminants that may be in untreated water include microorganisms such as viruses, protozoa and bacteria; inorganic contaminants such as salts and metals; organic chemical contaminants from industrial processes and petroleum use; pesticides and herbicides; and radioactive contaminants. Water quality depends on the local geology and ecosystem, as well as human uses such as sewage dispersion, industrial pollution, use of water bodies as a heat sink, and overuse (which may lower the level of the water).

The United States Environmental Protection Agency (EPA) limits the amounts of certain contaminants in tap water provided by US public water systems. The Safe Drinking Water Act authorizes EPA to issue two types of standards: *primary standards* regulate substances that potentially affect human health, and *secondary standards* prescribe aesthetic qualities, those that affect taste, odor, or appearance. The U.S. Food and Drug Administration (FDA) regulations establish limits for contaminants in bottled water that must provide the same protection for public health. Drinking water, including bottled water, may reasonably be expected to contain at least small amounts of some contaminants. The presence of these contaminants does not necessarily indicate that the water poses a health risk.

In urbanized areas around the world, water purification technology is used in municipal water systems to remove contaminants from the source water (surface water or groundwater) before it is distributed to homes, businesses, schools and other recipients. Water drawn directly from a stream, lake, or aquifer and that has no treatment will be of uncertain quality.

Industrial and Domestic Use

Dissolved minerals may affect suitability of water for a range of industrial and domestic purposes. The most familiar of these is probably the presence of ions of calcium and magnesium which interfere with the cleaning action of soap, and can form hard sulfate and soft carbonate deposits in water heaters or boilers. Hard water may be softened to remove these ions. The softening process often substitutes sodium cations. Hard water may be preferable to soft water for human consumption, since health problems have been associated with excess sodium and with calcium and magnesium deficiencies. Softening decreases nutrition and may increase cleaning effectiveness.

Environmental Water Quality

Urban runoff discharging to coastal waters

Environmental water quality, also called ambient water quality, relates to water bodies such as lakes, rivers, and oceans. Water quality standards for surface waters vary significantly due to different environmental conditions, ecosystems, and intended human uses. Toxic substances and high populations of certain microorganisms can present a health hazard for non-drinking purposes such as irrigation, swimming, fishing, rafting, boating, and industrial uses. These conditions may also affect wildlife, which use the water for drinking or as a habitat. Modern water quality laws generally specify protection of fisheries and recreational use and require, as a minimum, retention of current quality standards.

Satirical cartoon by William Heath, showing a woman observing monsters in a drop of London water (at the time of the *Commission on the London Water Supply* report, 1828)

There is some desire among the public to return water bodies to pristine, or pre-industrial conditions. Most current environmental laws focus on the designation of particular uses of a water body. In some countries these designations allow for some water contamination as long as the particular type of contamination is not harmful to the designated uses. Given the landscape changes (e.g., land development, urbanization, clearcutting in forested areas) in the watersheds of many freshwater bodies, returning to pristine conditions would be a significant challenge. In these cases, environmental scientists focus on achieving goals for maintaining healthy ecosystems and may concentrate on the protection of populations of endangered species and protecting human health.

Sampling and Measurement

The complexity of water quality as a subject is reflected in the many types of measurements of water quality indicators. The most accurate measurements of water quality are made on-site, because water exists in equilibrium with its surroundings. Measurements commonly made on-site and in direct contact with the water source in question include temperature, pH, dissolved oxygen, conductivity, oxygen reduction potential (ORP), turbidity, and Secchi disk depth.

Sample Collection

An automated sampling station installed along the East Branch Milwaukee River, New Fane, Wisconsin. The cover of the 24-bottle autosampler (center) is partially raised, showing the sample bottles inside. The autosampler was programmed to collect samples at time intervals, or proportionate to flow over a specified period. The data logger (white cabinet) recorded temperature, specific conductance, and dissolved oxygen levels.

More complex measurements are often made in a laboratory requiring a water sample to be collected, preserved, transported, and analyzed at another location. The process of water sampling introduces two significant problems. The first problem is the extent to which the sample may be representative of the water source of interest. Many water sources vary with time and with location. The measurement of interest may vary seasonally or from day to night or in response to some activity of man or natural populations of aquatic plants and animals. The measurement of interest may vary with distances from the water boundary with overlying atmosphere and underlying or confining soil. The sampler must determine if a single time and location meets the needs of the investigation, or if the water use of interest can be satisfactorily assessed by averaged values with time and/or location, or if critical maxima and minima require individual measurements over a range of times, locations and/or events. The sample collection procedure must assure correct weighting of individual sampling times and locations where averaging is appropriate. Where critical maximum or minimum values exist, statistical methods must be applied to observed variation to determine an adequate number of samples to assess probability of exceeding those critical values.

The second problem occurs as the sample is removed from the water source and begins to establish chemical equilibrium with its new surroundings - the sample container. Sample containers must be made of materials with minimal reactivity with substances to be measured; and pre-cleaning of sample containers is important. The water sample may dissolve part of the sample container and any residue on that container, or chemicals dissolved in the water sample may sorb onto the sample container and remain there when the water is poured out for analysis. Similar physical and chemical interactions may take place with any pumps, piping, or intermediate devices used to transfer the water sample into the sample container. Water collected from depths below the surface will normally be held at the reduced pressure of the atmosphere; so gas dissolved in the water may escape into unfilled space at the top of the container. Atmospheric gas present in that air space may also dissolve into the water sample. Other chemical reaction equilibria may change if the water sample changes temperature. Finely divided solid particles formerly suspended by water turbulence may settle to the bottom of the sample container, or a solid phase may form from biological growth or chemical precipitation. Microorganisms within the water sample may biochemically alter concentrations of oxygen, carbon dioxide, and organic compounds. Changing carbon dioxide concentrations may alter pH and change solubility of chemicals of interest. These problems are of special concern during measurement of chemicals assumed to be significant at very low concentrations.

Filtering a manually collected water sample (grab sample) for analysis

Sample preservation may partially resolve the second problem. A common procedure is keeping samples cold to slow the rate of chemical reactions and phase change, and analyzing the sample as soon as possible; but this merely minimizes the changes rather than preventing them. A useful procedure for determining influence of sample containers during delay between sample collection and analysis involves preparation for two artificial samples in advance of the sampling event. One sample container is filled with water known from previous analysis to contain no detectable amount of the chemical of interest. This sample, called a "blank," is opened for exposure to the atmosphere when the sample of interest is collected, then resealed and transported to the laboratory with the sample for analysis to determine if sample holding procedures introduced any measurable amount of the chemical of interest. The second artificial sample is collected with the sample of interest, but then "spiked" with a measured additional amount of the chemical of interest at the time of collection. The blank and spiked samples are carried with the sample of interest and analyzed by the same methods at the same times to determine any changes indicating gains or losses during the elapsed time between collection and analysis.

Testing in Response to Natural Disasters and Other Emergencies

Inevitably after events such as earthquakes and tsunamis, there is an immediate response by the aid agencies as relief operations get underway to try and restore basic infrastructure and provide the basic fundamental items that are necessary for survival and subsequent recovery. Access to clean drinking water and adequate sanitation is a priority at times like this. The threat of disease increases hugely due to the large numbers of people living close together, often in squalid conditions, and without proper sanitation.

After a natural disaster, as far as water quality testing is concerned there are widespread views on the best course of action to take and a variety of methods can be employed. The key basic water quality parameters that need to be addressed in an emergency are bacteriological indicators of fecal contamination, free chlorine residual, pH, turbidity and possibly conductivity/total dissolved solids. There are a number of portable water test kits on the market widely used by aid and relief agencies for carrying out such testing.

After major natural disasters, a considerable length of time might pass before water quality returns to pre-disaster levels. For example, following the 2004 Indian Ocean Tsunami the Colombo-based International Water Management Institute (IWMI) monitored the effects of saltwater and concluded that the wells recovered to pre-tsunami drinking water quality one and a half years after the event. IWMI developed protocols for cleaning wells contaminated by saltwater; these were subsequently officially endorsed by the World Health Organization as part of its series of Emergency Guidelines.

Chemical Analysis

A gas chromatograph-
mass spectrometer measures pesticides and other organic pollutants

The simplest methods of chemical analysis are those measuring chemical elements without respect to their form. Elemental analysis for oxygen, as an example, would indicate a concentration of 890,000 milligrams per litre (mg/L) of water sample because water is made of oxygen. The method selected to measure dissolved oxygen should differentiate between diatomic oxygen and oxygen combined with other elements. The comparative simplicity of elemental analysis has produced a large amount of sample data and water quality criteria for elements sometimes identified as heavy metals. Water analysis for heavy metals must consider soil particles suspended in the water sample. These suspended soil particles may contain measurable amounts of metal. Although the particles are not dissolved in the water, they may be consumed by people drinking the water. Adding acid to a water sample to prevent loss of dissolved metals onto the sample container may dissolve more metals from suspended soil particles. Filtration of soil particles from the water sample before acid addition, however, may cause loss of dissolved metals onto the filter. The complexities of differentiating similar organic molecules are even more challenging.

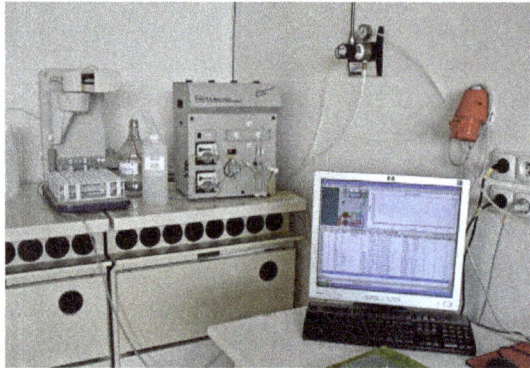

Atomic fluorescence spectroscopy is used to measure mercury and other heavy metals

Making these complex measurements can be expensive. Because direct measurements of water quality can be expensive, ongoing monitoring programs are typically conducted by government agencies. However, there are local volunteer programs and resources available for some general assessment. Tools available to the general public include on-site test kits, commonly used for home fish tanks, and biological assessment procedures.

Real-time Monitoring

Although water quality is usually sampled and analyzed at laboratories, nowadays, citizens demand real-time information about the water they are drinking. During the last years, several companies are deploying worldwide real-time remote monitoring systems for measuring water pH, turbidity or dissolved oxygen levels.

Drinking Water Indicators

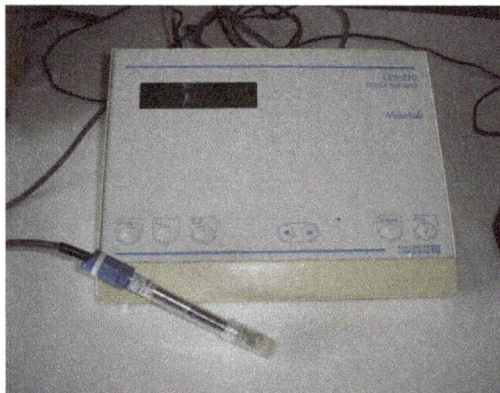

An electrical conductivity meter is used to measure total dissolved solids

The following is a list of indicators often measured by situational category:

- Alkalinity
- Color of water

- pH

- Taste and odor (geosmin, 2-Methylisoborneol (MIB), etc.)

- Dissolved metals and salts (sodium, chloride, potassium, calcium, manganese, magnesium)

- Microorganisms such as fecal coliform bacteria (*Escherichia coli*), Cryptosporidium, and Giardia lamblia;

- Dissolved metals and metalloids (lead, mercury, arsenic, etc.)

- Dissolved organics: colored dissolved organic matter (CDOM), dissolved organic carbon (DOC)

- Radon

- Heavy metals

- Pharmaceuticals

- Hormone analogs

Environmental Indicators

Physical Indicators

• Water Temperature • Specifics Conductance or EC, Electrical Conductance, Conductivity • Total suspended solids (TSS) • Transparency or Turbidity	• Total dissolved solids (TDS) • Odour of water • Color of water • Taste of water

Chemical Indicators

• pH • Biochemical oxygen demand (BOD) • Chemical oxygen demand (COD) • Dissolved oxygen (DO) • Total hardness (TH)	• Heavy metals • Nitrate • Orthophosphates • Pesticides • Surfactants

Biological Indicators

• Ephemeroptera	• *Escherichia coli* (E. coli)
• Plecoptera	
• Mollusca	• Coliform bacteria
• Trichoptera	

Biological monitoring metrics have been developed in many places, and one widely used measure is the presence and abundance of members of the insect orders Ephemeroptera, Plecoptera and Trichoptera. (Common names are, respectively, Mayfly, Stonefly and Caddisfly.) EPT indexes will naturally vary from region to region, but generally, within a region, the greater the number of taxa from these orders, the better the water quality. Organisations in the United States, such as EPA offer guidance on developing a monitoring program and identifying members of these and other aquatic insect orders.

Individuals interested in monitoring water quality who cannot afford or manage lab scale analysis can also use biological indicators to get a general reading of water quality. One example is the IOWATER volunteer water monitoring program, which includes a benthic macroinvertebrate indicator key.

Bivalve molluscs are largely used as bioindicators to monitor the health of aquatic environments in both fresh water and the marine environments. Their population status or structure, physiology, behaviour or the level of contamination with elements or compounds can indicate the state of contamination status of the ecosystem. They are particularly useful since they are sessile so that they are representative of the environment where they are sampled or placed. A typical project is the Mussel Watch Programme, but today they are used worldwide.

The Southern African Scoring System (SASS) method is a biological water quality monitoring system based on the presence of benthic macroinvertebrates. The SASS aquatic biomonitoring tool has been refined over the past 30 years and is now on the fifth version (SASS5) which has been specifically modified in accordance with international standards, namely the ISO/IEC 17025 protocol. The SASS5 method is used by the South African Department of Water Affairs as a standard method for River Health Assessment, which feeds the national River Health Programme and the national Rivers Database.

Standards and Reports

International

- The World Health Organisation (WHO) has published guidelines for drinking-water quality (GDWQ) in 2011.

- The International Organization for Standardization (ISO) published regulation of water quality in the section of ICS 13.060, ranging from water sampling, drinking water, industrial class water, sewage, and examination of water for chemical, physical or biological properties. ICS 91.140.60 covers the standards of water supply systems.

National Specifications for Ambient Water and Drinking Water

European Union

The water policy of the European Union is primarily codified in three directives:

- Directive on Urban Waste Water Treatment (91/271/EEC) of 21 May 1991 concerning discharges of municipal and some industrial wastewaters;

- The Drinking Water Directive (98/83/EC) of 3 November 1998 concerning potable water quality;

- Water Framework Directive (2000/60/EC) of 23 October 2000 concerning water resources management.

India

- Indian Council of Medical Research (ICMR) Standards for Drinking Water.

South Africa

Water quality guidelines for South Africa are grouped according to potential user types (e.g. domestic, industrial) in the 1996 Water Quality Guidelines. Drinking water quality is subject to the South African National Standard (SANS) 241 Drinking Water Specification.

United Kingdom

In England and Wales acceptable levels for drinking water supply are listed in the "Water Supply (Water Quality) Regulations 2000."

United States

In the United States, Water Quality Standards are defined by state agencies for various water bodies, guided by the desired uses for the water body (e.g., fish habitat, drinking water supply, recreational use). The Clean Water Act (CWA) requires each governing jurisdiction (states, territories, and covered tribal entities) to submit a set of biennial reports on the quality of water in their area. These reports are known as the 303(d) and 305(b) reports, named for their respective CWA provisions, and are submitted to, and approved by, EPA. These reports are completed by the governing jurisdiction, typically a state environmental agency. In coming years it is expected that the governing jurisdictions will submit all three reports as a single document, called the "Integrated Report." The *National Water Quality Inventory Report to Congress* is a general report on water quality, providing overall information about the number of miles of streams and rivers and their aggregate condition. The CWA requires states to adopt standards for each of the possible designated uses that they assign to their waters. Should evidence suggest or document that a stream, river or lake has failed to meet the water quality criteria for one or more of its designated uses, it is placed on a list of impaired waters. Once a state has placed a water body on this list, it must develop a management plan establishing Total Maximum Daily Loads for the pollutant(s) impairing the use of the water. These TMDLs establish the reductions needed to fully support the designated uses.

Drinking water standards, which are applicable to public water systems, are issued by EPA under the Safe Drinking Water Act.

References

- Linsley, Ray K. & Franzini, Joseph B. Water-Resources Engineering (1972) McGraw-Hill ISBN 0-07-037959-9 pp.454-456

- "Center for Coastal Monitoring and Assessment: Mussel Watch Contaminant Monitoring". Ccma.nos.noaa.gov. 2014-01-14. Retrieved 2015-09-04.

- Ogden, F. L.; Lai, W.; Steinke, R.C.; Zhu, J.; Talbot, C.A.; Wilson, J.L. (2015). "A new general 1-D vadose zone flow solution method". Water Resour. Res. 51: 4282. Bibcode:2015WRR....51.4282O. doi:10.1002/2015WR017126.

- Nouri, Hamideh; Beecham, Simon; Anderson, Sharoyn; Hassanli, Morad; Kazemi, Fatemeh (13 May 2014). "Remote sensing techniques for predicting evapotranspiration from mixed vegetated surfaces". Urban Water J. doi:10.1080/1573062X.2014.900092.

- Barati R, Akbari GH and Rahimi S (2013) Flood routing of an unmanaged river basin using Muskingum–Cunge model; field application and numerical experiments. Caspian Journal of Applied Sciences Research, 2(6):08-20.

- Nouri, Hamideh; Beecham, Simon; Kazemi, Fatemeh; Hassanli, Ali Morad (2013). "A review of ET measurement techniques for estimating the water requirements of urban landscape vegetation". Urban Water J. 10 (4): 247–259. doi:10.1080/1573062X.2012.726360.

- Akbari G. H, Barati R (2012). Comprehensive analysis of flooding in unmanaged catchments. Proceedings of the Institution of Civil Engineers-Water Management, 165(4): 229-238.

Tools and Techniques Used in Hydrology

The tools used in hydrology measure features of rainwater and groundwater with respect to pressure, quality, heat transfer, infiltration, properties, etc. The tools considered closely include piezometer, infiltrometer and rain gauge while the techniques discussed are aquifer test, Darcy's law, Bowen's ratio, and watertable control. The text also examines the hydrological transport model.

Piezometer

A piezometer is either a device used to measure liquid pressure in a system by measuring the height to which a column of the liquid rises against gravity, or a device which measures the pressure (more precisely, the piezometric head) of groundwater at a specific point. A piezometer is designed to measure static pressures, and thus differs from a pitot tube by not being pointed into the fluid flow.

Observation wells give some information on the water level in a formation, but must be read manually. Electrical pressure transducers of several types can be read automatically, making data acquisition more convenient.

Groundwater Measurement

Above-ground casing of a piezometer

The first piezometers in geotechnical engineering were open wells or standpipes (sometimes called Casagrande piezometers) installed into an aquifer. A Casagrande piezometer will typically have a solid casing down to the depth of interest, and a slotted or screened casing within the zone where water pressure is being measured. The casing is sealed into the drillhole with clay, bentonite or concrete to prevent surface water from contaminating the groundwater supply. In an unconfined aquifer, the water level in the piezometer would not be exactly coincident with the water table, especially when the vertical component of flow velocity is significant. In a confined aquifer under artesian conditions, the water level in the piezometer indicates the pressure in the aquifer, but not necessarily the water table. Piezometer wells can be much smaller in diameter than production wells, and a 5 cm diameter standpipe is common.

Piezometers in durable casings can be buried or pushed into the ground to measure the groundwater pressure at the point of installation. The pressure gauges (transducer) can be vibrating-wire, pneumatic, or strain-gauge in operation, converting pressure into an electrical signal. These piezometers are cabled to the surface where they can be read by data loggers or portable readout units, allowing faster or more frequent reading than is possible with open standpipe piezometers.

Infiltrometer

Infiltrometer is a device used to measure the rate of water infiltration into soil or other porous media. Commonly used infiltrometers are single ring or double ring infiltrometer, and also disc permeameter.

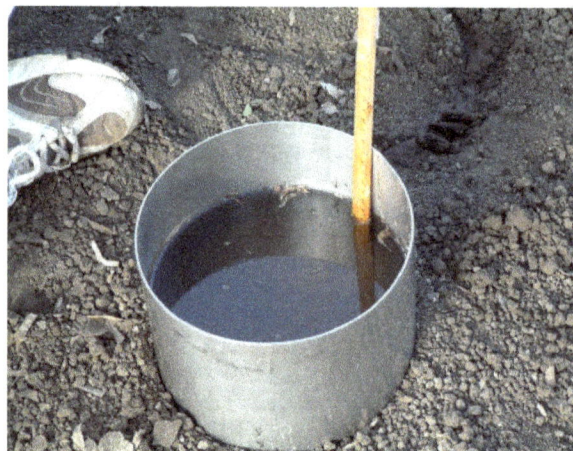

Single ring infiltrometer

The single ring involves driving a ring into the soil and supplying water in the ring either at constant head or falling head condition. Constant head refers to condition where the amount of water in the ring is always held constant. Because infiltration capacity is the maximum infiltration rate, and if infiltration rate exceeds the infiltration capacity, runoff will be the consequence, therefore maintaining constant head means the rate of water supplied corresponds to the infiltration capacity. The supplying of water is done with a Mariotte's bottle. Falling head refers to condition where water is supplied in the ring, and the water is allowed to drop with time. The operator records how much water goes into the soil for a given time period. The rate of which water goes into the soil is related to the soil's hydraulic conductivity.

Double ring infiltrometer requires two rings: an inner and outer ring. The purpose is to create a one-dimensional flow of water from the inner ring, as the analysis of data is simplified. If water is flowing in one-dimension at steady state condition, and a unit gradient is present in the underlying soil, the infiltration rate is approximately equal to the saturated hydraulic conductivity. An inner ring is driven into the ground, and a second bigger ring around that to help control the flow of water through the first ring. Water is supplied either with a constant or falling head condition, and the operator records how much water infiltrates from the inner ring into the soil over a given time period.

Double ring infiltrometer

There are three main problems related to the use of infiltrometers: 1. The pounding of the infiltrometer into the ground deforms the soil causing cracks and increasing the measured infiltration capacity. 2. Natural rainfall reaches terminal velocity. Also natural droplet sizes differ with different types of storms. Pouring water from a measuring cup however loses this momentum and variance. 3. With single ring infiltrometers, water spreads laterally as well as vertically and the analysis is more difficult.

Aquifer Test

An aquifer test (or a pumping test) is conducted to evaluate an aquifer by "stimulating" the aquifer through constant pumping, and observing the aquifer's "response" (drawdown) in observation wells. Aquifer testing is a common tool that hydrogeologists use to characterize a system of aquifers, aquitards and flow system boundaries.

A slug test is a variation on the typical aquifer test where an instantaneous change (increase or decrease) is made, and the effects are observed in the same well. This is often used in geotechnical or engineering settings to get a quick estimate (minutes instead of days) of the aquifer properties immediately around the well.

Aquifer tests are typically interpreted by using an analytical model of aquifer flow (the most fundamental being the Theis solution) to match the data observed in the real world, then assuming that the parameters from the idealized model apply to the real-world aquifer. In more complex cases, a numerical model may be used to analyze the results of an aquifer test, but adding complexity does not ensure better results.

Aquifer testing differs from well testing in that the behaviour of the well is primarily of concern in the latter, while the characteristics of the aquifer are quantified in the former. Aquifer testing also often utilizes one or more monitoring wells, or piezometers ("point" observation wells). A monitoring well is simply a well which is not being pumped (but is used to monitor the hydraulic head in the aquifer). Typically monitoring and pumping wells are screened across the same aquifers.

General Characteristics

Most commonly an aquifer test is conducted by pumping water from one well at a steady rate and for at least one day, while carefully measuring the water levels in the monitoring wells. When water is pumped from the pumping well the pressure in the aquifer that feeds that well declines. This decline in pressure will show up as drawdown (change in hydraulic head) in an observation well. Drawdown decreases with radial distance from the pumping well and drawdown increases with the length of time that the pumping continues.

The aquifer characteristics which are evaluated by most aquifer tests are:

- Hydraulic conductivity The rate of flow of water through a unit cross sectional area of an aquifer, at a unit hydraulic gradient. In US units the rate of flow is in gallons per day per square foot of cross sectional area; in SI units hydraulic conductivity is usually quoted in m^3 per day per m^2. Units are frequently shortened to metres per day or equivalent.

- Specific storage or storativity: a measure of the amount of water a confined aquifer will give up for a certain change in head;

- Transmissivity The rate at which water is transmitted through whole thickness and unit width of an aquifer under a unit hydraulic gradient. It is equal to the hydraulic conductivity times the thickness of an aquifer;

Additional aquifer characteristics which are sometimes evaluated, depending on the type of aquifer, include:

- Specific yield or drainable porosity: a measure of the amount of water an unconfined aquifer will give up when completely drained;

- Leakage coefficient: some aquifers are bounded by aquitards which slowly give up water to the aquifer, providing additional water to reduce drawdown;

- The presence of aquifer boundaries (recharge or no-flow) and their distance from the pumped well and piezometers.

Analysis Methods

An appropriate model or solution to the groundwater flow equation must be chosen to fit to the observed data. There are many different choices of models, depending on what factors are deemed important including:

- leaky aquitards,

- unconfined flow (delayed yield),

- partial penetration of the pumping and monitoring wells,

- finite wellbore radius — which can lead to wellbore storage,

- dual porosity (typically in fractured rock),

- anisotropic aquifers,

- heterogeneous aquifers,

- finite aquifers (the effects of physical boundaries are seen in the test), and

- combinations of the above situations.

Nearly all aquifer test solution methods are based on the Theis solution; it is built upon the most simplifying assumptions. Other methods relax one or more of the assumptions the Theis solution is built on, and therefore they get a more flexible (and more complex) result.

Transient Theis Solution

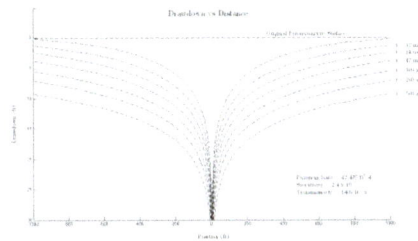

Cross-sectional plot of transient Theis solution for radial distance vs drawdown over time

The Theis equation was created by Charles Vernon Theis (working for the US Geological Survey) in 1935, from heat transfer literature (with the mathematical help of C.I. Lubin), for two-dimensional radial flow to a point source in an infinite, homogeneous aquifer. It is simply

$$s = \frac{Q}{4\pi T} W(u)$$

$$u = \frac{r^2 S}{4Tt}$$

where s is the drawdown (change in hydraulic head at a point since the beginning of the test), u is a dimensionless time parameter, Q is the discharge (pumping) rate of the well (volume divided by time, or m³/s), T and S are the transmissivity and storativity of the aquifer around the well (m²/s and unitless, respectively), r is the distance from the pumping well to the point where the drawdown was observed (m), t is the time since pumping began (seconds), and $W(u)$ is the "Well function" (called the exponential integral, E_1, in non-hydrogeology literature). The well function is approximated by the infinite series

$$W(u) = -0.577216 - \ln(u) + u - \frac{u^2}{2 \times 2!} + \frac{u^3}{3 \times 3!} - \frac{u^4}{4 \times 4!} + \cdots$$

Typically this equation is used to find the average T and S values near a pumping well, from drawdown data collected during an aquifer test. This is a simple form of inverse modeling, since the result (s) is measured in the well, r, t, and Q are observed, and values of T and S which best reproduce the measured data are put into the equation until a best fit between the observed data and the analytic solution is found.

The Theis solution is based on the following assumptions:

- The flow in the aquifer is adequately described by Darcy's law (i.e. Re<10).

- homogeneous, isotropic, confined aquifer,

- well is fully penetrating (open to the entire thickness (b) of aquifer),

- the well has zero radius (it is approximated as a vertical line) — therefore no water can be stored in the well,

- the well has a constant pumping rate Q,

- the head loss over the well screen is negligible,

- aquifer is infinite in radial extent,

- horizontal (not sloping), flat, impermeable (non-leaky) top and bottom boundaries of aquifer,

- groundwater flow is horizontal

- no other wells or long term changes in regional water levels (all changes in potentiometric surface are the result of the pumping well alone)

Even though these assumptions are rarely all met, depending on the degree to which they are violated (e.g., if the boundaries of the aquifer are well beyond the part of the aquifer which will be tested by the pumping test) the solution may still be useful.

Steady-state Thiem Solution

Steady-state radial flow to a pumping well is commonly called the Thiem solution, it comes about from application of Darcy's law to cylindrical shell control volumes (i.e., a cylinder with a larger radius which has a smaller radius cylinder cut out of it) about the pumping well; it is commonly written as:

$$h_0 - h = \frac{Q}{2\pi T} \ln\left(\frac{R}{r}\right)$$

In this expression h_o is the background hydraulic head, h_o-h is the drawdown at the radial distance r from the pumping well, Q is the discharge rate of the pumping well (at the origin), T is the transmissivity, and R is the radius of influence, or the distance at which the head is still h_o. These conditions (steady-state flow to a pumping well with no nearby boundaries) *never truly occur* in nature, but it can often be used as an approximation to actual conditions; the solution is derived by assuming there is a circular constant head boundary (e.g., a lake or river in full contact with the aquifer) surrounding the pumping well at a distance R.

Sources of Error

Of critical importance in both aquifer and well testing is the accurate recording of data. Not only must water levels and the time of the measurement be carefully recorded, but the pumping rates must be periodically checked and recorded. An unrecorded change in pumping rate of as little as 2% can be misleading when the data are analysed.

Bowen Ratio

In meteorology and hydrology, the Bowen ratio is used to describe the type of heat transfer in a water body. Heat transfer can either occur as sensible heat (differences in temperature without evapotranspiration) or latent heat (the energy required during a change of state, without a change in temperature). The Bowen ratio is the mathematical method generally used to calculate heat lost (or gained) in a substance; it is the ratio of energy fluxes from one state to another by sensible heat and latent heating respectively. It is calculated by the equation:

$$B = \frac{Q_h}{Q_e},$$

where Q_h is sensible heating and Q_e is latent heating. The quantity was named by Harald Sverdrup after Ira Sprague Bowen (1898–1973), an astrophysicist whose theoretical work on evaporation to air from water bodies made first use of it, and it is used most commonly in meteorology and hydrology. In this context, when the magnitude of B is less than one, a greater proportion of the available energy at the surface is passed to the atmosphere as latent heat than as sensible heat, and the converse is true for values of B greater than one. As $Q_e \to 0$, however, B becomes unbounded making the Bowen ratio a poor choice of variable for use in formulae, especially for arid surfaces. For this reason the evaporative fraction is sometimes a more appropriate choice of variable representing the relative contributions of the turbulent energy fluxes to the surface energy budget.

The Bowen ratio is related to the evaporative fraction, EF, through the equation,

$$EF = \frac{Q_e}{Q_e + Q_h} = \frac{1}{1+B}.$$

The Bowen ratio is an indicator of the type of surface. The Bowen ratio, B, is less than one over surfaces with abundant water supplies.

Type of surface	Range of Bowen ratios
Deserts	>10.0
Semi-arid landscapes	2.0-6.0
Temperate forests and grasslands	0.4-0.8
Tropical rainforests	0.1-0.3
Tropical oceans	<0.1

Darcy's Law

Darcy's law is a constitutive equation that describes the flow of a fluid through a porous medium. The law was formulated by Henry Darcy based on the results of experiments on the flow of water through beds of sand, forming the basis of hydrogeology, a branch of earth sciences.

Background

Although Darcy's law (an expression of Newton's second law) was determined experimentally by Darcy, it has since been derived from the Navier-Stokes equations via homogenization . It is analogous to Fourier's law in the field of heat conduction, Ohm's law in the field of electrical networks, or Fick's law in diffusion theory.

One application of Darcy's law is to analyze water flow through an aquifer; Darcy's law along with the equation of conservation of mass are equivalent to the groundwater flow equation, one of the basic relationships of hydrogeology. Darcy's law is also used to describe oil, water, and gas flows through petroleum reservoirs.

Description

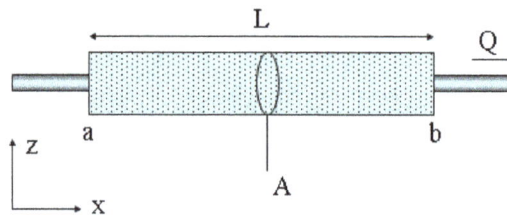

Diagram showing definitions and directions for Darcy's law.

Darcy's law, as refined by Morris Muskat, at constant elevation is a simple proportional relationship between the instantaneous discharge rate through a porous medium, the viscosity of the fluid and the pressure drop over a given distance.

$$Q = \frac{-\kappa A(p_b - p_a)}{\mu L}.$$

The total discharge, Q (units of volume per time, e.g., m³/s) is equal to the product of the intrinsic permeability of the medium, κ (m²), the cross-sectional area to flow, A (units of area, e.g., m²), and the total pressure drop $(p_b - p_a)$, (pascals), all divided by the viscosity, μ (Pa·s) and the length over which the pressure drop is taking place (L). The negative sign is needed because fluid flows from high pressure to low pressure. Note: the elevation head must be taken into account if the inlet and outlet are at different elevations. If the change in pressure is negative (where $p_a > p_b$), then the flow will be in the positive 'x' direction. Dividing both sides of the equation by the area and using more general notation leads

$$q = \frac{-\kappa}{\mu} \nabla p,$$

where q is the flux (discharge per unit area, with units of length per time, m/s) and ∇p is the pressure gradient vector (Pa/m). This value of flux, often referred to as the Darcy flux, is not the velocity which the fluid traveling through the pores is experiencing. The fluid velocity (v) is related to the Darcy flux (q) by the porosity (ϕ). The flux is divided by porosity to account for the fact that only a fraction of the total formation volume is available for flow. The fluid velocity would be the velocity a conservative tracer would experience if carried by the fluid through the formation.

$$v = \frac{q}{\phi}.$$

Darcy's law is a simple mathematical statement which neatly summarizes several familiar properties that groundwater flowing in aquifers exhibits, including:

- if there is no pressure gradient over a distance, no flow occurs (these are hydrostatic conditions),

- if there is a pressure gradient, flow will occur from high pressure towards low pressure (opposite the direction of increasing gradient - hence the negative sign in Darcy's law),

- the greater the pressure gradient (through the same formation material), the greater the discharge rate, and

- the discharge rate of fluid will often be different — through different formation materials (or even through the same material, in a different direction) — even if the same pressure gradient exists in both cases.

A graphical illustration of the use of the steady-state groundwater flow equation (based on Darcy's law and the conservation of mass) is in the construction of flownets, to quantify the amount of groundwater flowing under a dam.

Darcy's law is only valid for slow, viscous flow; fortunately, most groundwater flow cases fall in this category. Typically any flow with a Reynolds number less than one is clearly laminar, and it would be valid to apply Darcy's law. Experimental tests have shown that flow regimes with Reynolds numbers up to 10 may still be Darcian, as in the case of groundwater flow. The Reynolds number (a dimensionless parameter) for porous media flow is typically expressed as

$$Re = \frac{\rho v d_{30}}{\mu},$$

where ρ is the density of water (units of mass per volume), v is the specific discharge (not the pore velocity — with units of length per time), d_{30} is a representative grain diameter for the porous media (often taken as the 30% passing size from a grain size analysis using sieves - with units of length), and μ is the viscosity of the fluid.

Derivation

For stationary, creeping, incompressible flow, i.e. $D(\rho u_i)/Dt \approx 0$, the Navier-Stokes equation simplify to the Stokes equation:

$$\mu \nabla^2 u_i + \rho g_i - \partial_i p = 0,$$

where μ is the viscosity, u_i is the velocity in the i direction, g_i is the gravity component in the i direction and p is the pressure. Assuming the viscous resisting force is linear with the velocity we may write:

$$-\left(\kappa_{ij}\right)^{-1} \mu \phi u_j + \rho g_i - \partial_i p = 0,$$

where ϕ is the porosity, and κ_{ij} is the second order permeability tensor. This gives the velocity in the n direction,

$$\kappa_{ni}\left(\kappa_{ij}\right)^{-1} u_j = \delta_{nj} u_j = u_n = -\frac{\kappa_{ni}}{\phi \mu}\left(\partial_i p - \rho g_i\right),$$

which gives Darcy's law for the volumetric flux density in the n direction,

$$q_n = -\frac{\kappa_{ni}}{\mu}\left(\partial_i p - \rho g_i\right)..$$

In isotropic porous media the off-diagonal elements in the permeability tensor are zero, $\kappa_{ij} = 0$ for $i \neq j$ and the diagonal elements are identical, $\kappa = \kappa_{ii},$, and the common form is obtained

$$\mathbf{q} = -\frac{\kappa}{\mu}\left(\nabla p - \rho \mathbf{g}\right).$$

Additional Forms of Darcy's Law

Darcy's Law for Short Time Scales

For very short time scales, a time derivative of flux may be added to Darcy's law, which results in valid solutions at very small times (in heat transfer, this is called the modified form of Fourier's law),

$$\tau \frac{\partial q}{\partial t} + q = -\kappa \nabla h,,$$

where τ is a very small time constant which causes this equation to reduce to the normal form of Darcy's law at "normal" times (> nanoseconds). The main reason for doing this is that the regular groundwater flow equation (diffusion equation) leads to singularities at constant head boundaries at very small times. This form is more mathematically rigorous, but leads to a hyperbolic groundwater flow equation, which is more difficult to solve and is only useful at very small times, typically out of the realm of practical use.

Brinkman Form of Darcy's Law

Another extension to the traditional form of Darcy's law is the Brinkman term, which is used to account for transitional flow between boundaries (introduced by Brinkman in 1949),

$$\beta \nabla^2 q + q = \frac{-\kappa}{\mu} \nabla p,$$

where β is an effective viscosity term. This correction term accounts for flow through medium where the grains of the media are porous themselves, but is difficult to use, and is typically neglected.

Darcy's Law in Petroleum Engineering

Another derivation of Darcy's law is used extensively in petroleum engineering to determine the flow through permeable media - the most simple of which is for a one-dimensional, homogeneous rock formation with a fluid of constant viscosity.

$$Q = \frac{\kappa A}{\mu} \left(\frac{\partial p}{\partial x} \right),$$

where Q is the flowrate of the formation (in units of volume per unit time), k is the permeability of the formation (typically in millidarcies), A is the cross-sectional area of the formation, μ is the viscosity of the fluid (typically in units of centipoise. $\partial p / \partial x$ represents the pressure change per unit length of the formation. This equation can also be solved for permeability and is used to measure it, forcing a fluid of known viscosity through a core of a known length and area, and measuring the pressure drop across the length of the core.

Darcy-Forchheimer Law

For flows in porous media with Reynolds numbers greater than about 1 to 10, inertial effects can also become significant. Sometimes an inertial term is added to the Darcy's equation, known as Forchheimer term. This term is able to account for the non-linear behavior of the pressure difference vs flow data.

$$\frac{\partial p}{\partial x} = -\frac{\mu}{\kappa} q - \frac{\rho}{\kappa_1} q^2,$$

where the additional term κ_1 is known as inertial permeability.

Darcy's Law for Gases in Fine Media (Knudsen Diffusion or Klinkenberg Effect)

For gas flow in small characteristic dimensions (e.g., very fine sand, nanoporous structures etc.), the particle-wall interactions become more frequent, giving rise to additional wall friction (Knudsen friction). For a flow in this region, where both viscous and Knudsen friction are present, a new formulation needs to be used. Knudsen presented a semi-empirical model for flow in transition regime based on his experiments on small capillaries. For a porous media, the Knudsen equation can be given as

$$N = -\left(\frac{\kappa}{\mu}\frac{p_a + p_b}{2} + D_k^{\text{eff}}\right)\frac{1}{R_g T}\frac{p_b - p_a}{L},$$

where N is the molar flux, R_g is the gas constant, T is the temperature, D_k^{eff} is the effective Knudsen diffusivity of the porous media. The model can also be derived from first principles based binary friction model (BFM) . The differential equation of transition flow in porous media based on BFM is given as

$$\frac{\partial p}{\partial x} = -R_g T\left(\frac{\kappa p}{\mu} + D_K\right)^{-1} N.$$

This equation is valid for capillaries as well as porous media. The terminology of Knudsen effect and Knudsen diffusivity is more common in Mechanical and Chemical Engineering. In geological and petrochemical engineering, this effect is known as Klinkenberg effect. Using the definition of molar flux, the above equation can be rewritten as

$$\frac{\partial p}{\partial x} = -R_g T\left(\frac{\kappa p}{\mu} + D_K\right)^{-1} \frac{p}{R_g T} q.$$

This equation can be rearranged into the following equation

$$q = -\frac{\kappa}{\mu}\left(1 + \frac{D_k \mu}{\kappa}\frac{1}{p}\right)\frac{\partial p}{\partial x}.$$

Comparing this equation with conventional Darcy's law, a new formulation can be given as

$$q = -\frac{\kappa^{\text{eff}}}{\mu}\frac{\partial p}{\partial x}, \text{ where } \kappa^{\text{eff}} = \kappa\left(1 + \frac{D_k \mu}{\kappa}\frac{1}{p}\right).$$

This is equivalent to the effective permeability formulation proposed by Klinkenberg

$$\kappa^{\text{eff}} = \kappa\left(1 + \frac{b}{p}\right),$$

where b is known as the Klinkenberg parameter, which depends on the gas and the porous medium structure. This is quite evident if we compares the above formulations. The Klinkerberg parameter b is dependent on permeability, Knudsen diffusivity and viscosity (i.e., both gas and porous medium properties).

Validity of Darcy's Law

Darcy's Law is valid for laminar flow through sediments. In fine grained sediments, the dimensions of interstices are small and thus flow is laminar. Coarse-grained sediments also behave similarly but in very coarse-grained sediments the flow may be turbulent. Hence Darcy's Law is not always valid in such sediments. For flow through commercial pipes, the flow is laminar when Reynolds number is less than 2000, but in some sediments it has been found that flow is laminar when the value of Reynolds number is less than unity.

Hydrological Transport Model

River in Madagascar relatively free of sediment load

An hydrological transport model is a mathematical model used to simulate river or stream flow and calculate water quality parameters. These models generally came into use in the 1960s and 1970s when demand for numerical forecasting of water quality was driven by environmental legislation, and at a similar time widespread access to significant computer power became available. Much of the original model development took place in the United States and United Kingdom, but today these models are refined and used worldwide.

There are dozens of different transport models that can be generally grouped by pollutants addressed, complexity of pollutant sources, whether the model is steady state or dynamic, and time period modeled. Another important designation is whether the model is distributed (i.e. capable of predicting multiple points within a river) or lumped. In a basic model, for example, only one pollutant might be addressed from a simple point discharge into the receiving waters. In the most complex of models, various line source inputs from surface runoff might be added to multiple point sources, treating a variety of chemicals plus sediment in a dynamic environment including vertical river stratification and interactions of pollutants with in-stream biota. In addition watershed groundwater may also be included. The model is termed "physically based" if its parameters can be measured in the field.

Often models have separate modules to address individual steps in the simulation process. The most common module is a subroutine for calculation of surface runoff, allowing variation in land use type, topography, soil type, vegetative cover, precipitation and land management practice (such as the application rate of a fertilizer). The concept of hydrological modeling can be extended to other environments such as the oceans, but most commonly the subject of a river watershed is generally implied.

History

In 1850, T. J. Mulvany was probably the first investigator to use mathematical modeling in a stream hydrology context, although there was no chemistry involved. By 1892 M.E. Imbeau had conceived an event model to relate runoff to peak rainfall, again still with no chemistry. Robert E. Horton's seminal work on surface runoff along with his coupling of quantitative treatment of erosion laid the groundwork for modern chemical transport hydrology.

Types of Hydrological Transport Models

Physically Based Models

Physically based models (sometimes known as deterministic, comprehensive or process-based models) try to represent the physical processes observed in the real world. Typically, such models contain representations of surface runoff, subsurface flow, evapotranspiration, and channel flow, but they can be far more complicated. "Large scale simulation experiments were begun by the U.S. Army Corps of Engineers in 1953 for reservoir management on the main stem of the Missouri River". This, and other early work that dealt with the River Nile and the Columbia River are discussed, in a wider context, in a book published by the Harvard Water Resources Seminar, that contains the sentence just quoted. Another early model that integrated many submodels for basin chemical hydrology was the Stanford Watershed Model (SWM). The SWMM (Storm Water Management Model), the HSPF (Hydrological Simulation Program - FORTRAN) and other modern American derivatives are successors to this early work.

In Europe a favoured comprehensive model is the Système Hydrologique Européen (SHE), which has been succeeded by MIKE SHE and SHETRAN. MIKE SHE is a watershed-scale physically based, spatially distributed model for water flow and sediment transport. Flow and transport processes are represented by either finite difference representations of partial differential equations or by derived empirical equations. The following principal submodels are involved:

- Evapotranspiration: Penman-Monteith formalism

- Erosion: Detachment equations for raindrop and overland flow

- Overland and Channel Flow: Saint-Venant equations of continuity and momentum

- Overland Flow Sediment Transport: 2D total sediment load conservation equation

- Unsaturated Flow: Richards equation

- Saturated Flow: Darcy's law and the mass conservation of 2D laminar flow

- Channel Sediment Transport 1D mass conservation equation.

This model can analyze effects of land use and climate changes upon in-stream water quality, with consideration of groundwater interactions.

Worldwide a number of basin models have been developed, among them RORB (Australia), Xinanjiang (China), Tank model (Japan), ARNO (Italy), TOPMODEL (Europe), UBC (Canada) and HBV (Scandinavia), MOHID Land (Portugal). However, not all these models have a chemistry component. Generally speaking, SWM, SHE and TOPMODEL have the most comprehensive stream chemistry treatment and have evolved to accommodate the latest data sources including remote sensing and geographic information system data.

In the United States, the Corps of Engineers, Engineer Research and Development Center in conjunction with a researchers at a number of universities have developed the Gridded Surface/Subsurface Hydrologic Analysis GSSHA model. GSSHA is widely used in the U.S. for research and analysis by U.S. Army Corps of Engineers districts and larger consulting companies to compute flow, water levels, distributed erosion, and sediment delivery in complex engineering designs. A distributed nutrient and contaminant fate and transport component is undergoing testing. GSSHA input/output processing and interface with GIS is facilitated by the Watershed Modeling System (WMS).

Another model used in the United States and worldwide is V*flo*, a physics-based distributed hydrologic model developed by Vieux & Associates, Inc. V*flo* employs radar rainfall and GIS data to compute spatially distributed overland flow and channel flow. Evapotranspiration, inundation, infiltration, and snowmelt modeling capabilities are included. Applications include civil infrastructure operations and maintenance, stormwater prediction and emergency management, soil moisture monitoring, land use planning, water quality monitoring, and others.

Stochastic Models

These models based on data are black box systems, using mathematical and statistical concepts to link a certain input (for instance rainfall) to the model output (for instance runoff). Commonly used techniques are regression, transfer functions, neural networks and system identification. These models are known as stochastic hydrology models. Data based models have been used within hydrology to simulate the rainfall-runoff relationship, represent the impacts of antecedent moisture and perform real-time control on systems.

Model Components

Surface Runoff Modelling

A key component of a hydrological transport model is the surface runoff element, which allows assessment of sediment, fertilizer, pesticide and other chemical contaminants. Building on the work of Horton, the unit hydrograph theory was developed by Dooge in 1959. It required the presence of the National Environmental Policy Act and kindred other national legislation to provide the impetus to integrate water chemistry to hydrology model protocols. In the early 1970s the U.S. Environmental Protection Agency (EPA) began sponsoring a series of water quality models in response to the Clean Water Act. An example of these efforts was developed at the Southeast Water Laboratory, one of the first attempts to calibrate a surface runoff model with field data for a variety of chemical contaminants.

The attention given to surface runoff contaminant models has not matched the emphasis on pure hydrology models, in spite of their role in the generation of stream loading contaminant data. In the United States the EPA has had difficulty interpreting diverse proprietary contaminant models and has to develop its own models more often than conventional resource agencies, who, focused on flood forecasting, have had more of a centroid of common basin models.

Columbia River, which has surface runoff from agriculture and logging

Example Applications

Liden applied the HBV model to estimate the riverine transport of three different substances, nitrogen, phosphorus and suspended sediment in four different countries: Sweden, Estonia, Bolivia and Zimbabwe. The relation between internal hydrological model variables and nutrient transport was assessed. A model for nitrogen sources was developed and analysed in comparison with a statistical method. A model for suspended sediment transport in tropical and semi-arid regions was developed and tested. It was shown that riverine total nitrogen could be well simulated in the Nordic climate and riverine suspended sediment load could be estimated fairly well in tropical and semi-arid climates. The HBV model for material transport generally estimated material transport loads well. The main conclusion of the study was that the HBV model can be used to predict material transport on the scale of the drainage basin during stationary conditions, but cannot be easily generalised to areas not specifically calibrated. In a different work, Castanedo et al. applied an evolutionary algorithm to automated watershed model calibration.

Lake Tahoe, headwater sub-basin of the Truckee River watershed

The United States EPA developed the DSSAM Model to analyze water quality impacts from land use and wastewater management decisions in the Truckee River basin, an area which include the cities of Reno and Sparks, Nevada as well as the Lake Tahoe basin. The model satisfactorily predicted nutrient, sediment and dissolved oxygen parameters in the river. It is based on a pollutant loading metric called "Total Daily Maximum Load" (TDML). The success of this model contributed to the EPA's commitment to the use of the underlying TDML protocol in EPA's national policy for management of many river systems in the United States.

The DSSAM Model is constructed to allow dynamic decay of most pollutants; for example, total nitrogen and phosphorus are allowed to be consumed by benthic algae in each time step, and the algal communities are given a separate population dynamic in each river reach (e.g. based upon river temperature). Regarding stormwater runoff in Washoe County, the specific elements within a new xeriscape ordinance were analyzed for efficacy using the model. For the varied agricultural uses in the watershed, the model was run to understand the principal sources of impact, and management practices were developed to reduce in-river pollution. Use of the model has specifically been conducted to analyze survival of two endangered species found in the Truckee River and Pyramid Lake: the Cui-ui sucker fish (endangered 1967) and the Lahontan cutthroat trout (threatened 1970).

Watertable Control

Watertable control is the practice of controlling the height of the water table by drainage. Its main applications are in agricultural land (to improve the crop yield using agricultural drainage systems) and in cities to manage the extensive underground infrastructure that includes the foundations of large buildings, underground transit systems, and extensive utilities (water supply networks, sewerage, storm drains, and underground electrical grids).

Description and Definitions

Subsurface land drainage aims at controlling the water table of the groundwater in originally waterlogged land at a depth acceptable for the purpose for which the land is used. The depth of the water table with drainage is *greater* than without.

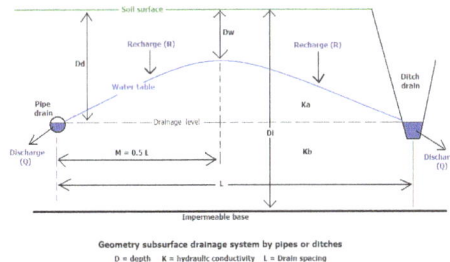

Figure 1. Drainage parameters in watertable control

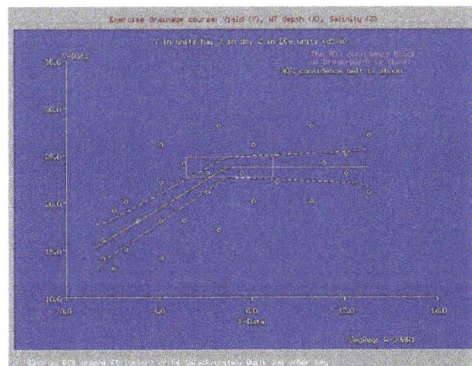

Figure 2. Crop yield (Y) and depth of water table (X in dm)

Purpose

In agricultural land drainage, the purpose of water table control is to establish a depth of the water table (Figure 1) that does no longer interfere negatively with the necessary farm operations and crop yields (Figure 2, made with the SegReg model).In addition, land drainage can help with soil salinity control.The soil's hydraulic conductivity plays an important role in drainage design.

The development of *agricultural drainage criteria* is required to give the designer and manager of the drainage system a target to achieve in terms of maintenance of an optimum depth of the water table.

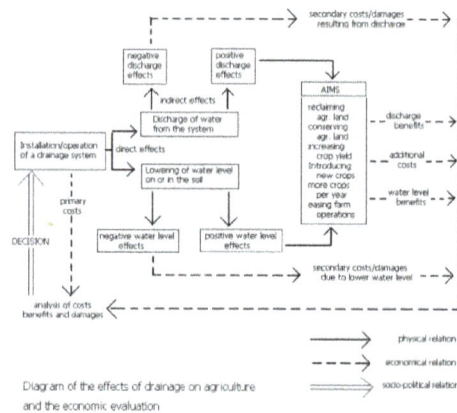

Figure 3. Positive and negative effects of land drainage

Optimization

Optimization of the depth of the water table is related the benefits and costs of the drainage system (Figure 3). The shallower the permissible depth of the water table, the lower the cost of the drainage system to be installed to achieve this depth. However, the lowering of the originally too shallow depth by land drainage entails *side effects*. These have also to be taken into account, including the costs of mitigation of negative side effects.

Drain depth (D_d, m), soil salinity (C_r, dS/m),
field Irrigation efficiency of the group A crops (FaA, -),
field irrigation sufficiency of the group A crops (JsA, -),
seasonal average depth of the water table (D_w, m), and
quantity of drainage water (G_d, mm per season).

Drain Depth	1st	season	(summer)		
D_d	C_r	FaA	JsA	D_w	G_d
0.6	2.7	0.84	0.99	0.37	105
0.8	2.5	0.83	0.98	0.55	112
1.0	2.4	0.82	0.97	0.74	117
1.2	2.2	0.81	0.96	0.93	122
1.4	2.1	0.80	0.95	1.12	127
	2nd	season	(winter)		
0.6	2.8	0.86	0.97	0.55	31
0.8	2.7	0.84	0.95	0.74	37
1.0	2.5	0.82	0.93	0.94	45
1.2	2.3	0.81	0.92	1.12	54
1.4	2.2	0.80	0.91	1.31	57

Figure 4. Example of effects of drain depth

Figure 4 shows an example of the effect of drain depth on soil salinity and various irrigation/drainage parameters as simulated by the SaltMod program.

History

Historically, agricultural land drainage started with the digging of relatively shallow open ditches that received both runoff from the land surface and outflow of groundwater. Hence the ditches had a surface as well as a subsurface drainage function.By the end of the 19th century and early in the 20th century it was felt that the ditches were a hindrance for the farm operations and the ditches were replaced by buried lines of clay pipes (tiles), each tile about 30 cm long. Hence the term "tile drainage". Since 1960, one started using long, flexible, corrugated plastic (PVC or PE) pipes that could be installed efficiently in one go by trenching machines. The pipes could be pre-wrapped with an envelope material, like synthetic fibre and geotextile, that would prevent the entry of soil particles into the drains.Thus, land drainage became a powerful industry. At the same time agriculture was steering towards maximum productivity, so that the installation of drainage systems came in full swing.

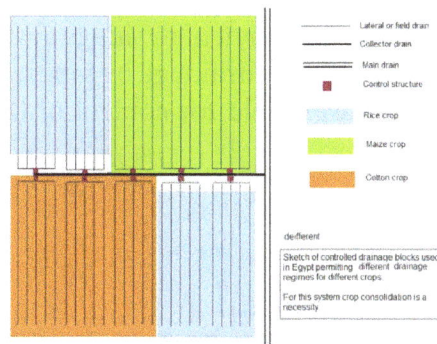

Figure 5. Controlled drainage

Environment

As a result of large scale developments, many modern drainage projects were *over-designed*, while the negative environmental side effects were ignored. In circles with environmental concern, the profession of land drainage got a poor reputation, sometimes justly so, sometimes unjustified, notably when land drainage was confused with the more encompassing activity of wetland reclamation. Nowadays, in some countries, the hardliner trend is reversed. Further, *checked* or *controlled* drainage systems were introduced, as shown in Figure 5 and discussed on the page: Drainage system (agriculture).

Drainage Design

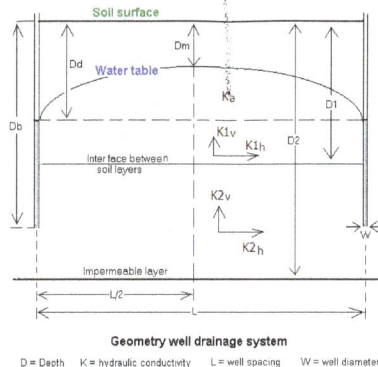

Figure 6. Geometry of a well drainage system

The design of subsurface drainage systems in terms layout, depth and spacing of the drains is often done using subsurface drainage equations with parameters like drain depth, depth of the water table, soil depth, hydraulic conductivity of the soil and drain discharge. The drain discharge is found from an agricultural water balance.The computations can be done using computer models like EnDrain.

Drainage by Wells

Subsurface drainage of groundwater can also be accomplished by pumped wells (*vertical* drainage, in contrast to *horizontal* drainage). Drainage wells have been used extensively in the Salinity Control and Reclamation Program (SCARP) in the Indus valley of Pakistan. Although the experiences were not overly successful, the feasibility of this technique in areas with deep and permeable aquifers is not to be discarded. The well spacings in these areas can be so wide (more than 1000m) that the installation of *vertical* drainage systems could be relatively cheap compared to *horizontal* subsurface drainage (drainage by pipes, ditches, trenches, at a spacing of 100m or less). For the design of a well field for control of the water table, the WellDrain model may be helpful.

Water Balance

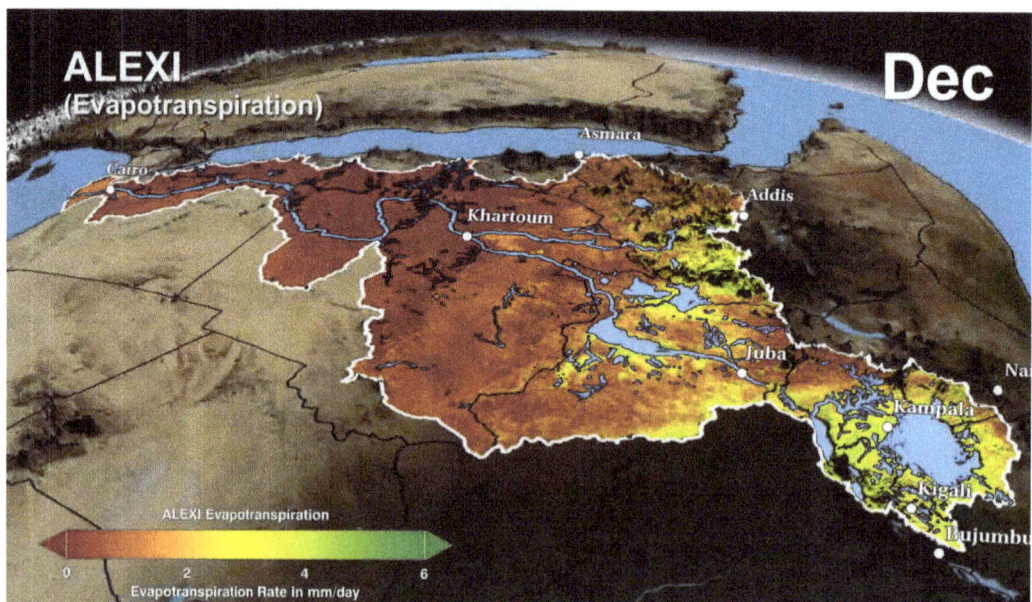

Nile basin water balance.

In hydrology, a water balance equation can be used to describe the flow of water in and out of a system. A system can be one of several hydrological domains, such as a column of soil or a drainage basin.Water balance can also refer to the ways in which an organism maintains water in dry or hot conditions. It is often discussed in reference to plants or arthropods, which have a variety of water retention mechanisms, including a lipid waxy coating that has limited permeability.

Equation

A general water balance equation is:

$$P = R + E + \Delta S$$

where

P is precipitation

E is evapotranspiration

R is streamflow

ΔS is the change in storage (in soil or the bedrock / ground water)

This equation uses the principles of conservation of mass in a closed system, whereby any water entering a system (via precipitation), must be transferred into either evaporation, surface runoff (eventually reaching the channel and leaving in the form of river discharge), or stored in the ground. This equation requires the system to be closed, and where it isn't (for example when surface runoff contributes to a different basin), this must be taken into account.

Extensive water balances are discussed in agricultural hydrology.

A water balance can be used to help manage water supply and predict where there may be water shortages. It is also used in irrigation, runoff assessment (e.g. through the *RainOff* model), flood control and pollution control. Further it is used in the design of subsurface drainage systems which may be *horizontal* (i.e. using pipes, tile drains or ditches) or *vertical* (drainage by wells). To estimate the drainage requirement, the use of an hydrogeological water balance and a groundwater model (e.g. SahysMod) may be instrumental.

The water balance can be illustrated using a water balance graph which plots levels of precipitation and evapotranspiration often on a monthly scale.

Several monthly water balance models had been developed for several conditions and purposes. Monthly water balance models had been studied since the 1940s.

References

- Dunnicliff, John (1993) [1988]. Geotechnical Instrumentation for Monitoring Field Performance. Wiley-Interscience. p. 117. ISBN 0-471-00546-0.

- A short history of the British Rainfall Organisation by DE Pedgley, Sept 2002, published by The Royal Meteorological Society ISBN 0-948090-21-9

- Hong Kong; Great Britain. Foreign and Commonwealth Office (1974). Jardi report Workman Chen, Hong Kong (PDF). HMSO. Retrieved 23 October 2011.

Various Applications of Hydrology

Hydrology finds valuable application in flood forecasting, hydroelectricity, antecedent moisture, drainage system, runoff model and discharge. This chapter considers how the tools and techniques of hydrology can be applied to improve human life. Agriculture is one of the major fields where hydrology is utilized.

Hydrology (Agriculture)

Agricultural hydrology is the study of water balance components intervening in agricultural water management, especially in irrigation and drainage.

Water Balance Components

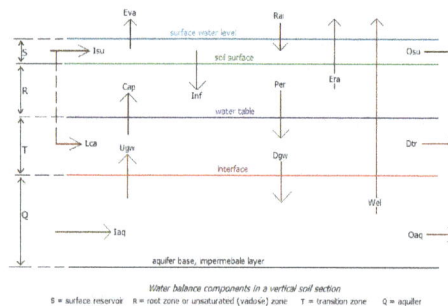

Water balance components in agricultural land

The water balance components can be grouped into components corresponding to zones in a vertical cross-section in the soil forming reservoirs with inflow, outflow and storage of water:

1. the surface reservoir (S)

2. the root zone or unsaturated (vadose zone) (R) with mainly vertical flows

3. the aquifer (Q) with mainly horizontal flows

4. a transition zone (T) in which vertical and horizontal flows are converted

The general water balance reads:

- inflow = outflow + change of storage

and it is applicable to each of the reservoirs or a combination thereof.

In the following balances it is assumed that the water table is inside the transition zone.

Surface Water Balance

The incoming water balance components into the surface reservoir (S) are:

1. Rai - Vertically incoming water to the surface e.g.: precipitation (including snow), rainfall, sprinkler irrigation

2. Isu - Horizontally incoming surface water. This can consist of natural inundation and/or surface irrigation

The outgoing water balance components from the surface reservoir (S) are:

1. Eva - Evaporation from open water on the soil surface

2. Osu - Surface runoff (natural) or surface drainage (artificial)

3. Inf - Infiltration of water through the soil surface into the root zone

The surface water balance reads:

- Rai + Isu = Eva + Inf + Osu + Ws, where Ws is the change of water storage on top of the soil surface

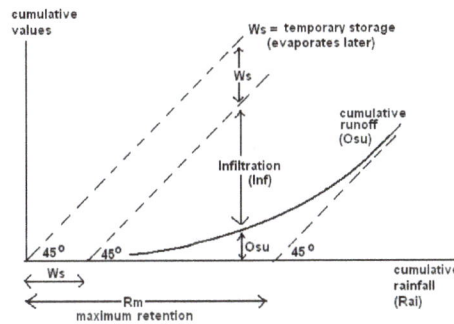

Principles of the Curve Number (CN) method

Surface runoff in the Curve Number method

Example of a surface water balance

An example is given of surface runoff according to the Curve number method. The applicable equation is:

- $Osu = (Rai - Ws)^2 / (Pp - Ws + Rm)$

where Rm is the *maximum retention* of the area for which the method is used

Normally one finds that Ws = 0.2 Rm and the value of Rm depends on the soil characteristics. The Curve Number method provides tables for these relations.

The method yields cumulative runoff values. To obtain runoff intensity values or runoff velocity (volume per unit of time) the cumulative duration is to be divided into sequential time steps (for example in hours).

Root Zone Water Balance

The incoming water balance components into the root zone (R) are:

1. Inf - Infiltration of water through the soil surface into the root zone

2. Cap - Capillary rise of water from the transition zone

The outgoing water balance components from the surface reservoir (R) are:

1. Era - Actual evaporation or evapotranspiration from the root zone

2. Per - Percolation of water from the unsaturated root zone into the transition zone

The root zone water balance reads:

*   Inf + Cap = Era + Per + Wr, where Wr is the change of water storage in the root zone

Transition Zone Water Balance

The incoming water balance components into the transition zone (T) are:

1. Per - Percolation of water from the unsaturated root zone into the transition zone

2. Lca - Infiltration of water from river, canal or drainage systems into the transition zone, often referred to as deep seepage losses

3. Ugw - Vertically upward seepage of water from the aquifer into the saturated transition zone

The outgoing water balance components from the transition zone (T) are:

1. Cap - Capillary rise of water into the root zone

2. Dtr - Artificial horizontal subsurface drainage

3. Dgw - Vertically downward drainage of water from the saturated transition zone into the aquifer

The water balance of the transition zone reads:

*   Per + Lca + Ugw = Cap + Dtr + Dgw + Wt, where Wt is the change of water storage in the transition zone noticeable as a change of the level of the water table.

Aquifer Water Balance

The incoming water balance components into the aquifer (Q) are:

1. Dgw - Vertically downward drainage of water from the saturated transition zone into the aquifer

2. Iaq - Horizontally incoming groundwater into the aquifer

The outgoing water balance components from the aquifer (Q) are:

1. Ugw - Vertically upward seepage of water from the aquifer into the saturated transition zone

2. Oaq - Horizontally outgoing groundwater from the aquifer

3. Wel - Discharge from (tube)wells placed in the aquifer

The water balance of the aquifer reads:

- Dgw + Iaq = Ugw + Wel + Oaq + Wq

where Wq is the change of water storage in the aquifer noticeable as a change of the artesian pressure.

Specific Water Balances

Combined Balances

Water balances can be made for a combination of two bordering vertical soil zones discerned, whereby the components constituting the inflow and outflow from one zone to the other will disappear. In long term water balances (month, season, year), the storage terms are often negligible small. Omitting these leads to *steady state* or *equilibrium* water balances.

Combination of surface reservoir (S)and root zone (R) in steady state yields the topsoil water Balance :

- Rai + Isu + Cap = Eva + Era + Osu + Per, where the linkage factor *Inf* has disappeared.

Combination of root zone (R) and transition zone (T) in steady state yields the subsoil water balance :

- Inf + Lca + Ugw = Era + Dtr + Dgw, where Wr the linkage factors *Per* and *Cap* have disappeared.

Combination of transition zone (T) and aquifer (Q) in steady state yields the geohydrologic water balance :

- Per + Lca + Iaq = Cap + Dtr + Wel + Oaq, where Wr the linkage factors *Ugw* and *Dgw* have disappeared.

Combining the uppermost three water balances in steady state gives the agronomic water balance :

- Rai + Isu + Lca + Ugw = Eva + Era + Osu + Dtr + Dgw, where the linkage factors *Inf, Per* and *Cap* have disappeared.

Combining all four water balances in steady state gives the overall water balance :

- Rai + Isu + Lca + Iaq = Eva + Era + Osu + Dtr + Wel + Oaq, where the linkage factors *Inf, Per, Cap, Ugw* and *Dgw* have disappeared.

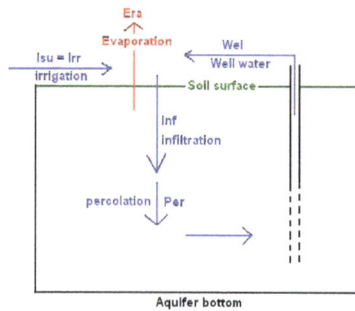

Reuse of percolation to the aquifer for irrigation
(Groundwater reuse)

Diagram for reuse of groundwater for irrigation by wells

Example of an overall water balance
An example is given of the reuse of groundwater for irrigation by pumped wells.

The total irrigation and the infiltration are:

• Inf = Irr + Wel, where *Irr* = surface irrigation from the canal system, and *Wel* = the irrigation from wells

The field irrigation efficiency (*Ff* < 1) is:

• Ff = Era / Inf, where *Era* = the evapotranspiration of the crop (consumptive use)

The value of *Era* is less than *Inf*, there is an excess of irrigation that percolates down to the subsoil (*Per*):

• Per = Irr + Wel − Era, or:

• Per = (1 - Ff) (Irr + Wel)

The percolation *Per* is pumped up again by wells for irrigation (*Wel*), hence:

• Wel = Per, or:

• Wel = (1 - Ff) (Irr + Wel), and therefore:

• Wel / Irr = (1 - Ff) / Ff

With this equation the following table can be prepared: |

Ff	0.20	0.25	0.33	0.50	0.75
Well / Irr	4	3	2	1	0.33

It can be seen that with low irrigation efficiency the amount of water pumped by the wells (*Wel*) is several time greater than the amount of irrigation water brought in by the canal system (*Irr*). This is due to the fact that a drop of water must be recirculated on the average several times before it is used by the plants.

Water Table Outside Transition Zone

When the water table is above the soil surface, the balances containing the components *Inf*, *Per*, *Cap* are not appropriate as they do not exist. When the water table is inside the root zone, the balances containing the components *Per*, *Cap* are not appropriate as they do not exist. When the water table is below the transition zone, only the *aquifer balance* is appropriate.

Reduced Number of Zones

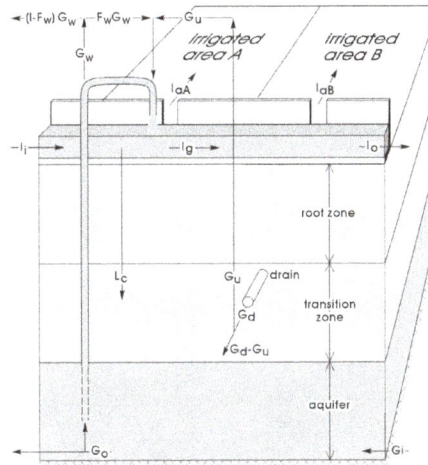

Saltmod water balance components

Under specific conditions it may be that no aquifer, transition zone and/or root zone is present. Water balances can be made omitting the absent zones.

Net and Excess Values

Vertical hydrological components along the boundary between two zones with arrows in the same direction can be combined into *net values* . For example : Npc = Per – Cap (net percolation), Ncp = Cap – Per (net capillary rise). Horizontal hydrological components in the same zone with arrows in same direction can be combined into *excess values* . For example : Egio = Iaq – Oaq (excess groundwater inflow over outflow), Egoi = Oaq – Iaq (excess groundwater outflow over inflow).

Salt Balances

Agricultural water balances are also used in the salt balances of irrigated lands. Further, the salt and water balances are used in agro-hydro-salinity-drainage models like Saltmod. Equally, they are used in groundwater salinity models like SahysMod which is a spatial variation of SaltMod using a polygonal network.

Irrigation and Drainage Requirements

The *irrigation requirement* (Irr) can be calculated from the *topsoil water balance*, the *agronomic water balance* and/or the *overall water balance*, as defined in the section "Combined balances", depending on the availability of data on the water balance components. Considering surface irrigation, assuming the evaporation of surface water is negligibly small (Eva = 0), setting the actual evapotranspiration Era equal to the potential evapotranspiration (Epo) so that Era = Epo and setting the surface inflow Isu equal to Irr so that Isu = Irr, the balances give respectively:

- Irr = Epo + Osu + Per – Rai – Cap

- $Irr = Epo + Osu + Dtr + Dgw - Rai - Lca - Ugw$

- $Irr = Epo + Osu + Dtr + Oaq - Rai - Lca - Iaq$

Defining the *irrigation efficiency* as $IEFF = Epo/Irr$, i.e. the fraction of the irrigation water that is consumed by the crop, it is found respectively that :

- $IEFF = 1 - (Osu + Per - Rai - Cap) / Irr$

- $IEFF = 1 - (Osu + Dtr + Dgw - Rai - Lca - Ugw) / Irr$

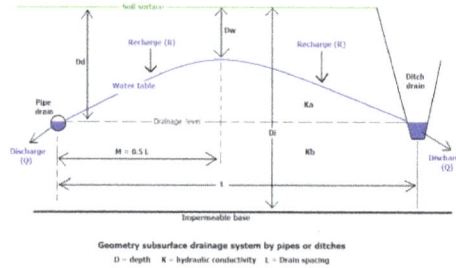

Geometry subsurface drainage system by pipes or ditches
D – depth   K – hydraulic conductivity   L – Drain spacing

The drain discharge determines the drain spacing

- $IEFF = 1 - (Osu + Dtr + Oaq - Rai - Lca - Iaq) / Irr$

Likewise the *safe yield* of wells, extracting water from the aquifer without overexploitation, can be determined using the *geohydrologic water balance* and/or the *overall water balance*, as defined in the section "Combined balances", depending on the availability of data on the water balance components.

Similarly, the subsurface drainage requirement can be found from the drain discharge (Dtr) in the *subsoil water balance*, the *agronomic water balance*, the *geohydrologic water balance* and/or the *overall water balance*.

In the same fashion, the well drainage requirement can be found from well discharge (Wel) in the *geohydrologic water balance* and/or the *overall water balance*.

The *subsurface drainage requirement* and *well drainage requirement* play an important role in the design of agricultural drainage systems (references:,).

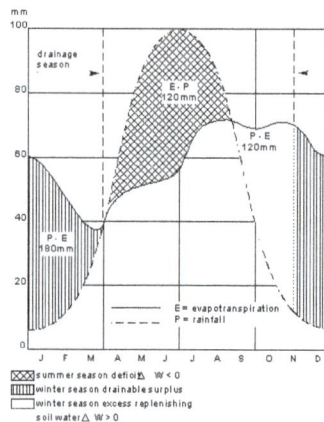

Average climatic data and drainage in the Netherlands

Example of drainage and irrigation requirements

The drainage and irrigation requirements in The Netherlands are derived from the climatic characteristics

Climatic data in the figure (mm)	Summer Apr-Aug	Winter Sep-Mar	Annual
Precipitation P	360	360	720
Evaporation E	480	60	540
Change of storage ΔW	−120	+120	0
Drainage requirement D	0	180	180
Irrigation requirement	variable	0	variable

The quantity of water to be drained in a normal winter is:

- $D = P - E - \Delta W$

According to the figure, the drainage period is from November to March (120 days) and the discharge of the drainage system is
$D = 180 / 120 = 1.5$ mm/day corresponding to 15 m³/day per ha.

During winters with more precipitation than normal, the drainage requirement increase accordingly.

The irrigation requirement depends on the rooting depth of the crops, which determines their capacity to make use of the water stored in the soil after winter. Having a shallow rooting system, pastures need irrigation to an amount of about half of the storage depletion in summer. Practically, wheat does not require irrigation because it develops deeper roots while during the maturing period a dry soil is favorable.

The analysis of cumulative frequency of climatic data plays an important role in the determination of the irrigation and drainage needs in the long run.

Flood Forecasting

Flood forecasting is the use of forecasted precipitation and streamflow data in rainfall-runoff and streamflow routing models to forecast flow rates and water levels for periods ranging from a few hours to days ahead, depending on the size of the watershed or river basin. Flood forecasting can also make use of forecasts of precipitation in an attempt to extend the lead-time available.

Flood forecasting is an important component of flood warning, where the distinction between the two is that the outcome of flood forecasting is a set of forecast time-profiles of channel flows or river levels at various locations, while "flood warning" is the task of making use of these forecasts to tell decisions on warnings of floods.

Hydroelectricity

Hydroelectricity is the term referring to electricity generated by hydropower; the production of electrical power through the use of the gravitational force of falling or flowing water. In 2015 hydropower generated 16.6% of the worlds total electricity and 70% of all renewable electricity, and is expected to increase about 3.1% each year for the next 25 years.

The Three Gorges Dam in Central China is the world's largest power producing facility of any kind.

Hydropower is produced in 150 countries, with the Asia-Pacific region generating 33 percent of global hydropower in 2013. China is the largest hydroelectricity producer, with 920 TWh of production in 2013, representing 16.9 percent of domestic electricity use.

The cost of hydroelectricity is relatively low, making it a competitive source of renewable electricity. The hydro station consumes no water, unlike coal or gas plants. The average cost of electricity from a hydro station larger than 10 megawatts is 3 to 5 U.S. cents per kilowatt-hour. With a dam and reservoir it is also a flexible source of electricity since the amount produced by the station can be changed up or down very quickly to adapt to changing energy demands. Once a hydroelectric complex is constructed, the project produces no direct waste, and has a considerably lower output level of greenhouse gases than fossil fuel powered energy plants.

History

Museum Hydroelectric power plant "Under the Town" in Serbia, built in 1900.

Hydropower has been used since ancient times to grind flour and perform other tasks. In the mid-1770s, French engineer Bernard Forest de Bélidor published *Architecture Hydraulique* which described vertical- and horizontal-axis hydraulic machines. By the late 19th century, the electrical generator was developed and could now be coupled with hydraulics. The growing demand for the Industrial Revolution would drive development as well. In 1878 the world's first hydroelectric power scheme was developed at Cragside in Northumberland, England by William George Armstrong. It was used to power a single arc lamp in his art gallery. The old Schoelkopf Power Station No. 1 near Niagara Falls in the U.S. side began to produce electricity in 1881. The first Edison hydroelectric power station, the Vulcan Street Plant, began operating September 30, 1882, in Appleton, Wisconsin, with an output of about 12.5 kilowatts. By 1886 there were 45 hydroelectric power stations in the U.S. and Canada. By 1889 there were 200 in the U.S. alone.

At the beginning of the 20th century, many small hydroelectric power stations were being constructed by commercial companies in mountains near metropolitan areas. Grenoble, France held the International Exhibition of Hydropower and Tourism with over one million visitors. By 1920 as 40% of the power produced in the United States was hydroelectric, the Federal Power Act was enacted into law. The Act created the Federal Power Commission to regulate hydroelectric power stations on federal land and water. As the power stations became larger, their associated dams developed additional purposes to include flood control, irrigation and navigation. Federal funding became necessary for large-scale development and federally owned corporations, such as the Tennessee Valley Authority (1933) and the Bonneville Power Administration (1937) were created. Additionally, the Bureau of Reclamation which had begun a series of western U.S. irrigation projects in the early 20th century was now constructing large hydroelectric projects such as the 1928 Hoover Dam. The U.S. Army Corps of Engineers was also involved in hydroelectric development, completing the Bonneville Dam in 1937 and being recognized by the Flood Control Act of 1936 as the premier federal flood control agency.

Hydroelectric power stations continued to become larger throughout the 20th century. Hydropower was referred to as *white coal* for its power and plenty. Hoover Dam's initial 1,345 MW power station was the world's largest hydroelectric power station in 1936; it was eclipsed by the 6809 MW Grand Coulee Dam in 1942. The Itaipu Dam opened in 1984 in South America as the largest, producing 14,000 MW but was surpassed in 2008 by the Three Gorges Dam in China at 22,500 MW. Hydroelectricity would eventually supply some countries, including Norway, Democratic Republic of the Congo, Paraguay and Brazil, with over 85% of their electricity. The United States currently has over 2,000 hydroelectric power stations that supply 6.4% of its total electrical production output, which is 49% of its renewable electricity.

Generating Methods

Turbine row at El Nihuil II Power Station in Mendoza, Argentina

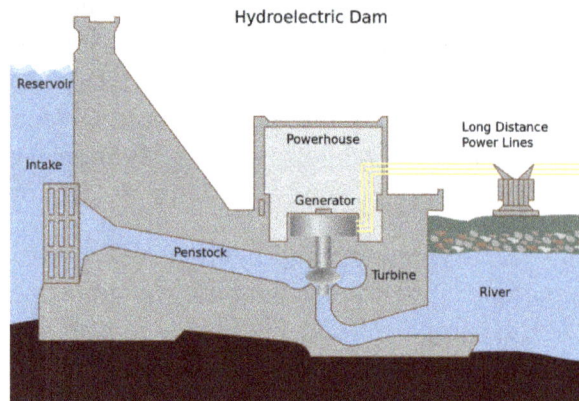

Cross section of a conventional hydroelectric dam.

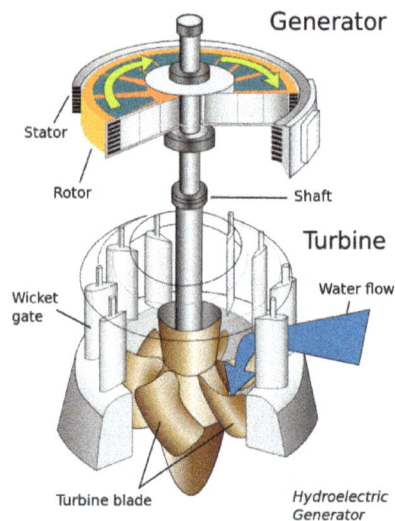

A typical turbine and generator

Conventional (dams)

Most hydroelectric power comes from the potential energy of dammed water driving a water turbine and generator. The power extracted from the water depends on the volume and on the difference in height between the source and the water's outflow. This height difference is called the head. A large pipe (the "penstock") delivers water from the reservoir to the turbine.

Pumped-storage

This method produces electricity to supply high peak demands by moving water between reservoirs at different elevations. At times of low electrical demand, the excess generation capacity is used to pump water into the higher reservoir. When the demand becomes greater, water is released back into the lower reservoir through a turbine. Pumped-storage schemes currently provide the most commercially important means of large-scale grid energy storage and improve the daily capacity factor of the generation system. Pumped storage is not an energy source, and appears as a negative number in listings.

Run-of-the-river

Run-of-the-river hydroelectric stations are those with small or no reservoir capacity, so that only the water coming from upstream is available for generation at that moment, and any oversupply must pass unused. A constant supply of water from a lake or existing reservoir upstream is a significant advantage in choosing sites for run-of-the-river. In the United States, run of the river hydropower could potentially provide 60,000 megawatts (80,000,000 hp) (about 13.7% of total use in 2011 if continuously available).

Tide

A tidal power station makes use of the daily rise and fall of ocean water due to tides; such sources are highly predictable, and if conditions permit construction of reservoirs, can also be dispatchable to generate power during high demand periods. Less common types of hydro schemes use water's kinetic energy or undammed sources such as undershot water wheels. Tidal power is viable in a relatively small number of locations around the world. In Great Britain, there are eight sites that could be developed, which have the potential to generate 20% of the electricity used in 2012.

Sizes, Types and Capacities of Hydroelectric Facilities

Large Facilities

Large-scale hydroelectric power stations are more commonly seen as the largest power producing facilities in the world, with some hydroelectric facilities capable of generating more than double the installed capacities of the current largest nuclear power stations.

Although no official definition exists for the capacity range of large hydroelectric power stations, facilities from over a few hundred megawatts are generally considered large hydroelectric facilities.

Currently, only four facilities over 10 GW (10,000 MW) are in operation worldwide, see table below.

Rank	Station	Country	Location	Capacity (MW)
1.	Three Gorges Dam	China	30°49′15″N 111°00′08″E30.82083°N 111.00222°E	22,500
2.	Itaipu Dam	Brazil Paraguay	25°24′31″S 54°35′21″W25.40861°S 54.58917°W	14,000
3.	Xiluodu Dam	China	28°15′35″N 103°38′58″E28.25972°N 103.64944°E	13,860
4.	Guri Dam	Venezuela	07°45′59″N 62°59′57″W7.76639°N 62.99917°W	10,200

Panoramic view of the Itaipu Dam, with the spillways (closed at the time of the photo) on the left. In 1994, the American Society of Civil Engineers elected the Itaipu Dam as one of the seven modern Wonders of the World.

Small

Small hydro is the development of hydroelectric power on a scale serving a small community or industrial plant. The definition of a small hydro project varies but a generating capacity of up to 10 megawatts (MW) is generally accepted as the upper limit of what can be termed small hydro. This may be stretched to 25 MW and 30 MW in Canada and the United States. Small-scale hydro-electricity production grew by 28% during 2008 from 2005, raising the total world small-hydro capacity to 85 GW. Over 70% of this was in China (65 GW), followed by Japan (3.5 GW), the United States (3 GW), and India (2 GW).

A micro-hydro facility in Vietnam

Pico hydroelectricity in Mondulkiri, Cambodia

Small hydro stations may be connected to conventional electrical distribution networks as a source of low-cost renewable energy. Alternatively, small hydro projects may be built in isolated areas that would be uneconomic to serve from a network, or in areas where there is no national electrical distribution network. Since small hydro projects usually have minimal reservoirs and civil construction work, they are seen as having a relatively low environmental impact compared to large hydro. This decreased environmental impact depends strongly on the balance between stream flow and power production.

Micro

Micro hydro is a term used for hydroelectric power installations that typically produce up to 100 kW of power. These installations can provide power to an isolated home or small community, or are sometimes connected to electric power networks. There are many of these installations around the world, particularly in developing nations as they can provide an economical source of energy without purchase of fuel. Micro hydro systems complement photovoltaic solar energy systems because in many areas, water flow, and thus available hydro power, is highest in the winter when solar energy is at a minimum.

Pico

Pico hydro is a term used for hydroelectric power generation of under 5 kW. It is useful in small, remote communities that require only a small amount of electricity. For example, to power one or two fluorescent light bulbs and a TV or radio for a few homes. Even smaller turbines of 200-300 W may power a single home in a developing country with a drop of only 1 m (3 ft). A Pico-hydro setup is typically run-of-the-river, meaning that dams are not used, but rather pipes divert some of the flow, drop this down a gradient, and through the turbine before returning it to the stream.

Underground

An underground power station is generally used at large facilities and makes use of a large natural height difference between two waterways, such as a waterfall or mountain lake. An underground tunnel is constructed to take water from the high reservoir to the generating hall built in an underground cavern near the lowest point of the water tunnel and a horizontal tailrace taking water away to the lower outlet waterway.

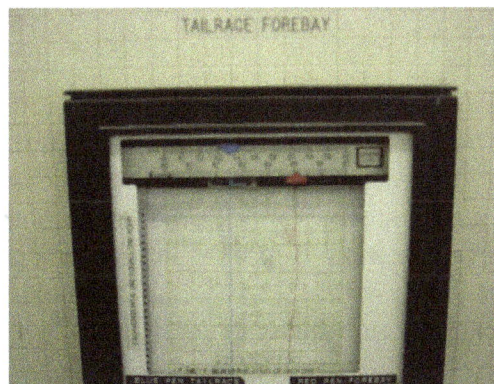

Measurement of the tailrace and forebay rates at the Limestone Generating Station in Manitoba, Canada.

Calculating Available Power

A simple formula for approximating electric power production at a hydroelectric station is: $P = \rho hrgk$, where

- *P* is Power in watts,

- ρ is the density of water (~1000 kg/m³),

- *h* is height in meters,

- *r* is flow rate in cubic meters per second,

- *g* is acceleration due to gravity of 9.8 m/s²,

- *k* is a coefficient of efficiency ranging from 0 to 1. Efficiency is often higher (that is, closer to 1) with larger and more modern turbines.

Annual electric energy production depends on the available water supply. In some installations, the water flow rate can vary by a factor of 10:1 over the course of a year.

Advantages and Disadvantages

Advantages

The Ffestiniog Power Station can generate 360 MW of electricity within 60 seconds of the demand arising.

Flexibility

Hydropower is a flexible source of electricity since stations can be ramped up and down very quickly to adapt to changing energy demands. Hydro turbines have a start-up time of the order of a few minutes. It takes around 60 to 90 seconds to bring a unit from cold start-up to full load; this is much shorter than for gas turbines or steam plants. Power generation can also be decreased quickly when there is a surplus power generation. Hence the limited capacity of hydropower units is not generally used to produce base power except for vacating the flood pool or meeting downstream needs. Instead, it serves as backup for non-hydro generators.

Low Power Costs

The major advantage of hydroelectricity is elimination of the cost of fuel. The cost of operating a hydroelectric station is nearly immune to increases in the cost of fossil fuels such as oil, natural gas or coal, and no imports are needed. The average cost of electricity from a hydro station larger than 10 megawatts is 3 to 5 U.S. cents per kilowatt-hour.

Hydroelectric stations have long economic lives, with some plants still in service after 50–100 years. Operating labor cost is also usually low, as plants are automated and have few personnel on site during normal operation.

Where a dam serves multiple purposes, a hydroelectric station may be added with relatively low construction cost, providing a useful revenue stream to offset the costs of dam operation. It has been calculated that the sale of electricity from the Three Gorges Dam will cover the construction costs after 5 to 8 years of full generation. Additionally, some data shows that in most countries large hydropower dams will be too costly and take too long to build to deliver a positive risk adjusted return, unless appropriate risk management measures are put in place.

Suitability for Industrial Applications

While many hydroelectric projects supply public electricity networks, some are created to serve specific industrial enterprises. Dedicated hydroelectric projects are often built to provide the substantial amounts of electricity needed for aluminium electrolytic plants, for example. The Grand Coulee Dam switched to support Alcoa aluminium in Bellingham, Washington, United States for American World War II airplanes before it was allowed to provide irrigation and power to citizens (in addition to aluminium power) after the war. In Suriname, the Brokopondo Reservoir was constructed to provide electricity for the Alcoa aluminium industry. New Zealand's Manapouri Power Station was constructed to supply electricity to the aluminium smelter at Tiwai Point.

Reduced CO_2 emissions

Since hydroelectric dams do not burn fossil fuels, they do not directly produce carbon dioxide. While some carbon dioxide is produced during manufacture and construction of the project, this is a tiny fraction of the operating emissions of equivalent fossil-fuel electricity generation. One measurement of greenhouse gas related and other externality comparison between energy sources can be found in the ExternE project by the Paul Scherrer Institute and the University of Stuttgart which was funded by the European Commission. According to that study, hydroelectricity produces the least amount of greenhouse gases and externality of any energy source. Coming in second place was wind, third was nuclear energy, and fourth was solar photovoltaic. The low greenhouse gas impact of hydroelectricity is found especially in temperate climates. The above study was for local energy in Europe; presumably similar conditions prevail in North America and Northern Asia, which all see a regular, natural freeze/thaw cycle (with associated seasonal plant decay and regrowth). Greater greenhouse gas emission impacts are found in the tropical regions because the reservoirs of power stations in tropical regions produce a larger amount of methane than those in temperate areas.

Other Uses of the Reservoir

Reservoirs created by hydroelectric schemes often provide facilities for water sports, and become tourist attractions themselves. In some countries, aquaculture in reservoirs is common. Multiuse dams installed for irrigation support agriculture with a relatively constant water supply. Large hydro dams can control floods, which would otherwise affect people living downstream of the project.

Disadvantages

Ecosystem Damage and Loss of Land

Hydroelectric power stations that use dams would submerge large areas of land due to the requirement of a reservoir. Merowe Dam in Sudan.

Large reservoirs associated with traditional hydroelectric power stations result in submersion of extensive areas upstream of the dams, sometimes destroying biologically rich and productive lowland and riverine valley forests, marshland and grasslands. Damming interrupts the flow of rivers and can harm local ecosystems, and building large dams and reservoirs often involves displacing people and wildlife. The loss of land is often exacerbated by habitat fragmentation of surrounding areas caused by the reservoir.

Hydroelectric projects can be disruptive to surrounding aquatic ecosystems both upstream and downstream of the plant site. Generation of hydroelectric power changes the downstream river environment. Water exiting a turbine usually contains very little suspended sediment, which can lead to scouring of river beds and loss of riverbanks. Since turbine gates are often opened intermittently, rapid or even daily fluctuations in river flow are observed.

Siltation and Flow Shortage

When water flows it has the ability to transport particles heavier than itself downstream. This has a negative effect on dams and subsequently their power stations, particularly those on rivers or within catchment areas with high siltation. Siltation can fill a reservoir and reduce its capacity to control floods along with causing additional horizontal pressure on the upstream portion of the dam. Eventually, some reservoirs can become full of sediment and useless or over-top during a flood and fail.

Changes in the amount of river flow will correlate with the amount of energy produced by a dam. Lower river flows will reduce the amount of live storage in a reservoir therefore reducing the amount of water that can be used for hydroelectricity. The result of diminished river flow can be power shortages in areas that depend heavily on hydroelectric power. The risk of flow shortage may increase as a result of climate change. One study from the Colorado River in the United States suggest that modest climate changes, such as an increase in temperature in 2 degree Celsius resulting in a 10% decline in precipitation, might reduce river run-off by up to 40%. Brazil in particular is vulnerable due to its heavy reliance on hydroelectricity, as increasing temperatures, lower water flow and alterations in the rainfall regime, could reduce total energy production by 7% annually by the end of the century.

Methane Emissions (from Reservoirs)

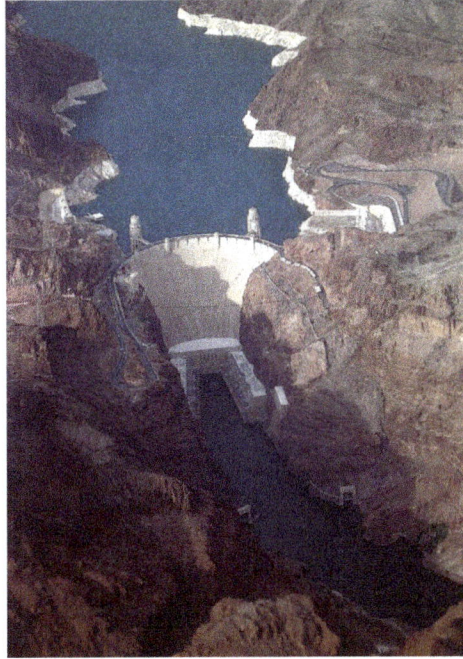

The Hoover Dam in the United States is a large conventional dammed-hydro facility, with an installed capacity of 2,080 MW.

Lower positive impacts are found in the tropical regions, as it has been noted that the reservoirs of power plants in tropical regions produce substantial amounts of methane. This is due to plant material in flooded areas decaying in an anaerobic environment, and forming methane, a greenhouse gas. According to the World Commission on Dams report, where the reservoir is large compared to the generating capacity (less than 100 watts per square metre of surface area) and no clearing of the forests in the area was undertaken prior to impoundment of the reservoir, greenhouse gas emissions from the reservoir may be higher than those of a conventional oil-fired thermal generation plant.

In boreal reservoirs of Canada and Northern Europe, however, greenhouse gas emissions are typically only 2% to 8% of any kind of conventional fossil-fuel thermal generation. A new class of underwater logging operation that targets drowned forests can mitigate the effect of forest decay.

Relocation

Another disadvantage of hydroelectric dams is the need to relocate the people living where the reservoirs are planned. In 2000, the World Commission on Dams estimated that dams had physically displaced 40-80 million people worldwide.

Failure Risks

Because large conventional dammed-hydro facilities hold back large volumes of water, a failure due to poor construction, natural disasters or sabotage can be catastrophic to downriver settlements and infrastructure. Dam failures have been some of the largest man-made disasters in history.

During Typhoon Nina in 1975 Banqiao Dam failed in Southern China when more than a year's worth of rain fell within 24 hours. The resulting flood resulted in the deaths of 26,000 people, and another 145,000 from epidemics. Millions were left homeless. Also, the creation of a dam in a geologically inappropriate location may cause disasters such as 1963 disaster at Vajont Dam in Italy, where almost 2,000 people died.

The Malpasset Dam failure in Fréjus on the French Riviera (Côte d'Azur), southern France, collapsed on December 2, 1959, killing 423 people in the resulting flood.

Smaller dams and micro hydro facilities create less risk, but can form continuing hazards even after being decommissioned. For example, the small Kelly Barnes Dam failed in 1967, causing 39 deaths with the Toccoa Flood, ten years after its power station was decommissioned the earthen embankment dam failed.

Comparison With other Methods of Power Generation

Hydroelectricity eliminates the flue gas emissions from fossil fuel combustion, including pollutants such as sulfur dioxide, nitric oxide, carbon monoxide, dust, and mercury in the coal. Hydroelectricity also avoids the hazards of coal mining and the indirect health effects of coal emissions. Compared to nuclear power, hydroelectricity construction requires altering large areas of the environment while a nuclear power station has a small footprint, and hydro-powerstation failures have caused tens of thousands of more deaths than any nuclear station failure. The creation of Garrison Dam, for example, required Native American land to create Lake Sakakawea, which has a shoreline of 1,320 miles, and caused the inhabitants to sell 94% of their arable land for $7.5 million in 1949.

Compared to wind farms, hydroelectricity power stations have a more predictable load factor. If the project has a storage reservoir, it can generate power when needed. Hydroelectric stations can be easily regulated to follow variations in power demand.

World Hydroelectric Capacity

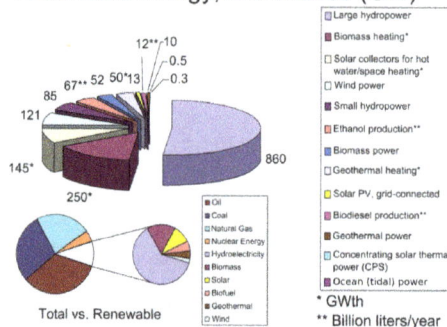

World renewable energy share (2008)

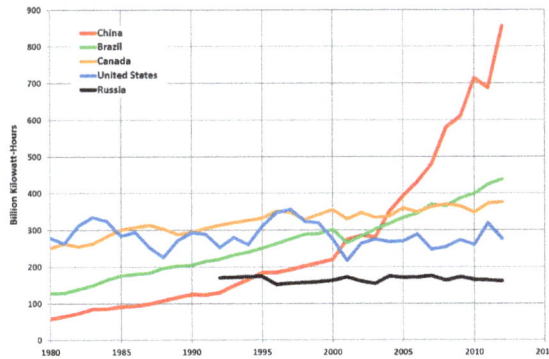

Trends in the top five hydroelectricity-producing countries

The ranking of hydro-electric capacity is either by actual annual energy production or by installed capacity power rating. In 2015 hydropower generated 16.6% of the worlds total electricity and 70% of all renewable electricity. Hydropower is produced in 150 countries, with the Asia-Pacific region generated 32 percent of global hydropower in 2010. China is the largest hydroelectricity producer, with 721 terawatt-hours of production in 2010, representing around 17 percent of domestic electricity use. Brazil, Canada, New Zealand, Norway, Paraguay, Austria, Switzerland, and Venezuela have a majority of the internal electric energy production from hydroelectric power. Paraguay produces 100% of its electricity from hydroelectric dams, and exports 90% of its production to Brazil and to Argentina. Norway produces 98–99% of its electricity from hydroelectric sources.

A hydro-electric station rarely operates at its full power rating over a full year; the ratio between annual average power and installed capacity rating is the capacity factor. The installed capacity is the sum of all generator nameplate power ratings.

Ten of the largest hydroelectric producers as at 2013.				
Country	Annual hydroelectric production (TWh)	Installed capacity (GW)	Capacity factor	% of total production
China	920	194	0.37	16.9%
Canada	392	76	0.59	60.1%
Brazil	391	86	0.56	68.6%
United States	290	102	0.42	6.7
Russia	183	50	0.42	17.3%
India	142	40	0.43	11.9%
Norway	129	31	0.49	96.1%
Japan	85	49	0.37	8.1%
Venezuela	84	15	0.67	67.8%
France	76	25	0.46	13.2%

Major Projects Under Construction

Name	Maximum Capacity	Country	Construction started	Scheduled completion	Comments
Belo Monte Dam	11,181 MW	Brazil	March, 2011	2015	Preliminary construction underway. Construction suspended 14 days by court order Aug 2012
Siang Upper HE Project	11,000 MW	India	April, 2009	2024	Multi-phase construction over a period of 15 years. Construction was delayed due to dispute with China.
Tasang Dam	7,110 MW	Burma	March, 2007	2022	Controversial 228 meter tall dam with capacity to produce 35,446 GWh annually.
Xiangjiaba Dam	6,400 MW	China	November 26, 2006	2015	The last generator was commissioned on July 9, 2014
Grand Ethiopian Renaissance Dam	6,000 MW	Ethiopia	2011	2017	Located in the upper Nile Basin, drawing complaint from Egypt
Nuozhadu Dam	5,850 MW	China	2006	2017	
Jinping 2 Hydropower Station	4,800 MW	China	January 30, 2007	2014	To build this dam, 23 families and 129 local residents need to be moved. It works with Jinping 1 Hydropower Station as a group.
Diamer-Bhasha Dam	4,500 MW	Pakistan	October 18, 2011	2023	
Jinping 1 Hydropower Station	3,600 MW	China	November 11, 2005	2014	The sixth and final generator was commissioned on 15 July 2014
Jirau Power Station	3,300 MW	Brazil	2008	2013	Construction halted in March 2011 due to worker riots.
Guanyinyan Dam	3,000 MW	China	2008	2015	Construction of the roads and spillway started.
Lianghekou Dam	3,000 MW	China	2014	2023	
Dagangshan Dam	2,600 MW	China	August 15, 2008	2016	
Liyuan Dam	2,400 MW	China	2008	2013	
Tocoma Dam Bolívar State	2,160 MW	Venezuela	2004	2014	This power station would be the last development in the Low Caroni Basin, bringing the total to six power stations on the same river, including the 10,000MW Guri Dam.
Ludila Dam	2,100 MW	China	2007	2015	Brief construction halt in 2009 for environmental assessment.
Shuangjiangkou Dam	2,000 MW	China	December, 2007	2018	The dam will be 312 m high.

Ahai Dam	2,000 MW	China	July 27, 2006	2015	
Teles Pires Dam	1,820 MW	Brazil	2011	2015	
Site C Dam	1,100 MW	Canada	2015	2024	First large dam in western Canada since 1984
Lower Subansiri Dam	2,000 MW	India	2007	2016	

Antecedent Moisture

Antecedent moisture is a term from the fields of hydrology and sewage collection and disposal that describes the relative wetness or dryness of a watershed or sanitary sewershed. Antecedent moisture conditions change continuously and can have a very significant effect on the flow responses in these systems during wet weather. The effect is evident in most hydrologic systems including stormwater runoff and sanitary sewers with inflow and infiltration. Many modeling and analysis challenges that are created by antecedent moisture conditions are evident within combined sewers and separate sanitary sewer systems.

Definition

The word antecedent simply means "preceding conditions". Combining the terms "antecedent" and "moisture" together means "preceding wetness conditions". Antecedent moisture is a term that describes the relative wetness or dryness of a sewershed, which changes continuously and can have a very significant effect on the flow responses in these systems during wet weather. Antecedent moisture conditions are high when there has been a lot of recent rainfall and the ground is moist. Antecedent moisture conditions are low when there has been little rainfall and the ground becomes dry.

Hydrologic Basis

Rainfall/runoff relationship are well defined within the field of hydrology. Surface runoff in hydrologic systems is generally conceptualized as occurring from pervious and impervious areas. It is the pervious runoff that is affected by antecedent moisture conditions, as runoff from impervious surfaces such as roads, sidewalks, and roofs will not be significantly affected by preceding moisture levels. Pervious surfaces (such as fields, woods, grassed areas, and open areas) are highly affected by antecedent moisture conditions, as they will produce a greater rate of runoff when they are wet than when they are dry.

Rainfall-dependent inflow and infiltration (RDII) into sewer systems is highly affected by antecedent moisture conditions, and these effects can be more complex than the rainfall/runoff relationships for surface water. The travel paths for RDII entering the sewer system are more complex than surface water runoff, because the transport mechanisms include both surface runoff and subsurface transportation. This adds additional complexities to the hydrologic effects and antecedent moisture effects such as the saturation levels of the soils in the subsurface, groundwater levels, and subsurface hydraulics.

Antecedent moisture conditions are highly affected by preceding rainfall levels. However, preceding rainfall is not the only condition that affects antecedent moisture, and many other variables in the hydrologic process can have a significant impact. For example, air temperature, wind speed, and humidity levels affect evaporation rates, which can significantly change antecedent moisture conditions. Additional effects may include evapotranspiration, presence or absence of tree canopy, and snow and ice melting effects.

Traditional Analysis Approaches

Traditional approaches for analyzing antecedent moisture effects rely on physically-based models derived from first principles, such as the principles of energy, momentum, and continuity, which rely on measurements of many parameters for input and simulation. These include programs such as the Storm Water Management Model, Mouse RDII, or other rainfall/runoff simulation programs. These models are frequently calibrated to a specific antecedent moisture condition observed during a single storm. Fitting data from several storms that occurred during various antecedent moisture conditions requires modifying the model parameters and recalibrating the model. At the end of this process, the modeler is left with several models, each of which can fit a specific storm that occurring during a specific antecedent moisture condition, but none of which are capable of simultaneously fitting all of the data. This is the challenge of using event-based models with traditional approaches: it requires the user to select a particular antecedent moisture condition for design simulations.

Some modeling approaches—such as the Hydrologic Simulation Program - Fortran (HSPF) or the Stanford Watershed Model developed by Crawford and Linsley (1966)—attempt to address antecedent moisture conditions through a complex physically based representation of the transport paths of water on the surface and in the subsurface. These tools have their place in researching and studying the various complexities associated with hydrologic transport processes. However, the great number of parameters in these models, the difficulty of measuring the many parameters, and the sensitivity of the model output to slight variations in the parameters makes using these models to simulate antecedent moisture in sewer systems challenging. Occam's razor provides evidence of these challenges from a systems perspective.

Data-based Approaches

An alternative approach for modeling antecedent moisture is to start from measurements of the behavior of the system and the external influences (inputs to the system) and try to determine a mathematical relation between them without going into the details of what is actually happening inside the system. This approach is called system identification. System identification is applied in several fields beyond engineering, ranging from economics to astronomy, and it also comes under other names (such as inverse modeling, time series analysis, and empirical physical modeling). System identification is a general term to describe mathematical tools and algorithms that build dynamical models from measured data. A dynamical model in this context is a mathematical description of the dynamic behavior of a system or process. In many cases, a so-called white-box model based on first principles (e.g., a model for a physical process from Newton's laws of motion) will be overly complex and possibly even impossible to obtain in reasonable time, due to the complex nature of many systems and processes.

Data-based approaches based on system identification, such as the i3D antecedent moisture model, have been applied to hydrologic modeling for simulating antecedent moisture effects on wet weather events in sanitary collection systems. This modeling approach differs from traditional techniques because it is based on system identification and is guided by system observations (i.e. data) and mathematical routines are used to generate the correct model structure, rather than physically based first principles. This is in contrast to assuming that the correct model is known beforehand, as is typically the case for modeling within civil engineering. This technique allows information within the observations to guide the modeling algorithms so that only the relevant and observed dynamics are present in the model structure. The resulting models are not black box, but are grey box models that have parameters and structure that tie directly to physical understanding and interpretation.

Drainage System (Agriculture)

An agricultural drainage system is a system by which water is drained on or in the soil to enhance agricultural production of crops. It may involve any combination of stormwater control, erosion control, and watertable control.

Classification

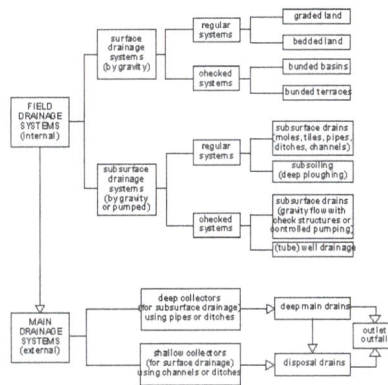

Classification of agricultural drainage systems.

Agriculture drainage is mainly of two types.. (1). Surface drainage (2). Sub Surface Drainage. However there are some other drainage systems also. Figure 1 classifies the various types of drainage systems. It shows the field (or internal) and the main (or external) systems. The function of the field drainage system is to control the water table, whereas the function of the main drainage system is to collect, transport, and dispose of the water through an outfall or outlet. In some instances one makes an additional distinction between collector and main drainage systems. Field drainage systems are differentiated in surface and subsurface field drainage systems. Sometimes (e.g. in irrigated, submerged rice fields), a form of temporary drainage is required whereby the drainage system is allowed to function on certain occasions only (e.g. during the harvest period). If allowed to function continuously, excessive quantities of water would be lost. Such a system is therefore called a checked, or controlled, drainage system. More usually, however, the drainage system is meant to function as regularly as possible to prevent undue waterlogging at any time and one employs a regular drainage system. In literature, this is sometimes also called a "relief drainage system".

Surface Drainage Systems

The regular surface drainage systems, which start functioning as soon as there is an excess of rainfall or irrigation applied, operate entirely by gravity. They consist of reshaped or reformed land surfaces and can be divided into:

- Bedded systems, used in flat lands for crops other than rice;

- Graded systems, used in sloping land for crops other than rice.

The bedded and graded systems may have ridges and furrows.m The checked surface drainage systems consist of check gates placed in the embankments surrounding flat basins, such as those used for rice fields in flat lands. These fields are usually submerged and only need to be drained on certain occasions (e.g. at harvest time). Checked surface drainage systems are also found in terraced lands used for rice. In literature, not much information can be found on the relations between the various regular surface field drainage systems, the reduction in the degree of waterlogging, and the agricultural or environmental effects. It is therefore difficult to develop sound agricultural criteria for the regular surface field drainage systems. Most of the known criteria for these systems concern the efficiency of the techniques of land leveling and earthmoving. Similarly, agricultural criteria for checked surface drainage systems are not very well known...

Subsurface Drainage Systems

Mug and sole drain (Scotland, 18th century)

Like the surface field drainage systems, the subsurface field drainage systems can also be differentiated in regular systems and checked (controlled) systems.

When the drain discharge takes place entirely by gravity, both types of subsurface systems have much in common, except that the checked systems have control gates that can be opened and closed according to need. They can save much irrigation water. A *checked* drainage system also reduces the discharge through the main drainage system, thereby reducing construction costs.

When the discharge takes place by *pumping*, the drainage can be checked simply by not operating the pumps or by reducing the pumping time. In northwestern India, this practice has increased the irrigation efficiency and reduced the quantity of irrigation water needed, and has not led to any undue salinization.

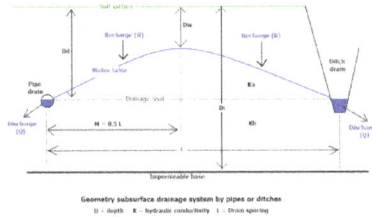

Parameters of horizontal drainage

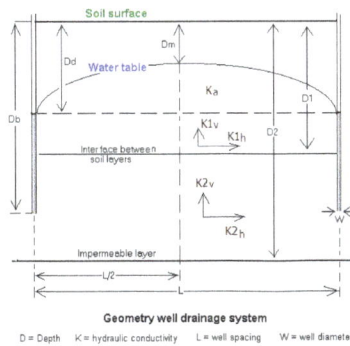

Parameters of vertical drainage

The subsurface field drainage systems consist of horizontal or slightly sloping channels made in the soil; they can be open ditches, trenches, filled with brushwood and a soil cap, filled with stones and a soil cap, buried pipe drains, tile drains, or mole drains, but they can also consist of a series of wells.

- Innovation of Century Hydroluis Drainage Pipe System. solving all problem of clogging in land fit period and after land period. the system is working by water filtration system working by Archimedes law. drainage pipe and his special cover on the top of drainage pipe giving the drainage pipe many advantages. Advantage & Characteristics of Hydroluis® Drainage ; First Anti-Plant roots Drainage pipe On the world. (It does not emit moisture from the pipe holes). Anti-bacterial (Iron Oxide was effecting Iron Ochre, Calcium carbonate and sulfate was effecting bio-film clogging problems on other drain pipe envelope). First drainage pipes saving the underground water in drought seasons. Works only when water table rises above specified levels. Best performance flowing and minimum sediment entrance than all system during land fit period. Eliminates requirement for annual maintenance or internal cleaning of drainage pipe and guarantees the strength life cycle and operational performance of plastic. Showing longer-term operating performance in all types of soil conditions as compared to competing drainage systems. Long term operating costs of the drainage pipe proves to be the most cost-effective. After laying the system, disadvantages of growing plant roots are turned in advantages as this system increases water flow in the direction of drainage pipe. Usable in shallow impermeable grounds,and ropy ground i.e., near the plant roots.

Modern buried pipe drains often consist of corrugated, flexible, and perforated plastic (PE or PVC) pipe lines wrapped with an *envelope* or filter material to improve the permeability around the pipes and to prevent entry of soil particles, which is especially important in fine sandy and silty soils. The surround may consist of synthetic fibre (geotextile).

The *field drains* (or *laterals*) discharge their water into the collector or main system either by *gravity* or by *pumping*. The wells (which may be open dug wells or *tubewells*) have normally to be pumped, but sometimes they are connected to drains for discharge by gravity.

Subsurface drainage by wells is often referred to as vertical drainage, and drainage by channels as horizontal drainage, but it is more clear to speak of "field drainage by wells" and "field drainage by ditches or pipes" respectively.

In some instances, subsurface drainage can be achieved simply by breaking up slowly permeable soil layers by deep plowing (sub-soiling), provided that the underground has sufficient natural drainage. In other instances, a combination of sub-soiling and subsurface drains may solve the problem.

Main Drainage Systems

Deep collector drain

The main drainage systems consist of deep or shallow collectors, and main drains or disposal drains.

Deep collector drains are required for subsurface field drainage systems, whereas shallow collector drains are used for surface field drainage systems, but they can also be used for pumped subsurface systems. The deep collectors may consist of open ditches or buried pipe lines.

The terms *deep collectors* and *shallow collectors* refer rather to the depth of the water level in the collector below the soil surface than to the depth of the bottom of the collector. The bottom depth is determined both by the depth of the water level and by the required discharge capacity.

The deep collectors may either discharge their water into deep main drains (which are drains that do not receive water directly from field drains, but only from collectors), or their water may be pumped into a disposal drain.

Disposal drains are main drains in which the depth of the water level below the soil surface is not bound to a minimum, and the water level may even be above the soil surface, provided that embankments are made to prevent inundation. Disposal drains can serve both subsurface and surface field drainage systems.

Pumping station Van Sasse in Grave, the Netherlands

Deep main drains can gradually become disposal drains if they are given a smaller gradient than the land slope along the drain.

The technical criteria applicable to main drainage systems depend on the hydrological situation and on the type of system.

Main Drainage Outlet

The final point of a main drainage system is the gravity outlet structure or the pumping station.

Applications

Surface drainage systems are usually applied in relatively flat lands that have soils with a low or medium infiltration capacity, or in lands with high-intensity rainfalls that exceed the normal infiltration capacity, so that frequent waterlogging occurs on the soil surface. Subsurface drainage systems are used when the drainage problem is mainly that of shallow water tables. When both surface and subsurface waterlogging occur, a combined surface/subsurface drainage system is required. Sometimes, a subsurface drainage system is installed in soils with a low infiltration capacity, where a surface drainage problem may improve the soil structure and the infiltration capacity so greatly that a surface drainage system is no longer required. On the other hand, it can also happen that a surface drainage system diminishes the recharge of the groundwater to such an extent that the subsurface drainage problem is considerably reduced or even eliminated. The choice between a subsurface drainage system by pipes and ditches or by tube wells is more a matter of technical criteria and costs than of agricultural criteria, because both types of systems can be designed to meet the same agricultural criteria and achieve the same benefits. Usually, pipe drains or ditches are preferable to wells. However, when the soil consists of a poorly permeable top layer several meters thick, overlying a rapidly permeable and deep subsoil, wells may be a better option, because the drain spacing required for pipes or ditches would be considerably smaller than the spacing for wells.

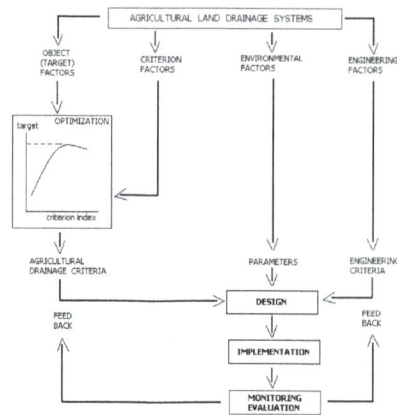

The role of agricultural, environmental, and engineering criteria in the
optimization, design, and evaluation of agricultural land drainage systems.

Drainage design procedures

When the land needs a subsurface drainage system, but saline groundwater is present at great depth, it is better to employ a shallow, closely spaced system of pipes or ditches instead of a deep, widely spaced system. The reason is that the deeper systems produce a more salty effluent than the shallow systems. Environmental criteria may then prohibit the use of the deeper systems. In some drainage projects, one may find that only main drainage systems are envisaged. The agricultural land is then still likely to suffer from field drainage problems. In other cases, one may find that field drainage systems are ineffective because there is no adequate main drainage system. In either case, the installation of drainage systems is not recommended. Reference: gives a general description of land drainage in the world and shows a paper on types of agricultural land drainage systems used in different parts of the world.

Drainage System Design

The analysis of positive and negative (side) effects of drainage and the optimization of drainage design in accordance to the *drainage design procedures*.

Runoff Model (Reservoir)

A runoff model is a mathematical model describing the rainfall–runoff relations of a rainfall *catchment area*, drainage basin or *watershed*. More precisely, it produces a surface runoff hydrograph in response to a rainfall event, represented by and input as a hyetograph. In other words, the model calculates the conversion of rainfall into runoff. A well known runoff model is the *linear reservoir*, but in practice it has limited applicability. The runoff model with a *non-linear reservoir* is more universally applicable, but still it holds only for catchments whose surface area is limited by the condition that the rainfall can be considered more or less uniformly distributed over the area. The maximum size of the watershed then depends on the rainfall characteristics of the region. When the study area is too large, it can be divided into sub-catchments and the various runoff hydrographs may be combined using flood routing techniques.

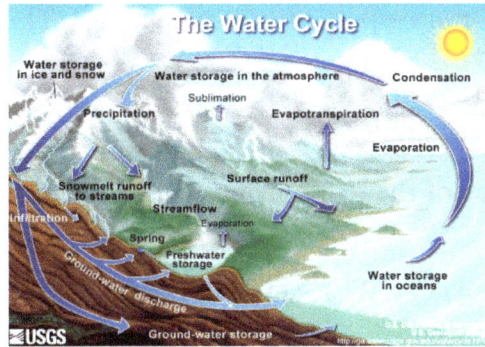

Rainfall-runoff models need to be calibrated before they can be used.

Linear Reservoir

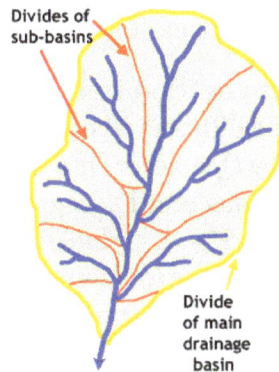

A watershed or drainage basin

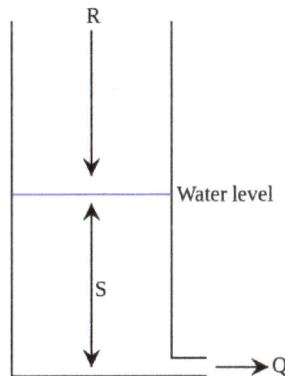

Figure 1. A linear reservoir

The hydrology of a linear reservoir (figure 1) is governed by two equations.

1. flow equation: $Q = A.S$, with units [L/T], where L is length (e.g. mm) and T is time (e.g. h, day)

2. continuity or water balance equation: $R = Q + dS/dT$, with units [L/T]

where:
Q is the *runoff* or *discharge*
R is the *effective rainfall* or *rainfall excess* or *recharge*

A is the constant *reaction factor* or *response factor* with unit [1/T]
S is the water storage with unit [L]
dS is a differential or small increment of S
dT is a differential or small increment of T

Runoff equation A combination of the two previous equations results in a differential equation, whose solution is:

- $Q_2 = Q_1 \exp\{-A(T_2 - T_1)\} + R[1 - \exp\{-A(T_2 - T_1)\}]$

This is the *runoff equation* or *discharge equation*, where Q_1 and Q_2 are the values of Q at time T_1 and T_2 respectively while T_2-T_1 is a small time step during which the recharge can be assumed constant.

Computing the total hydrograph Provided the value of A is known, the *total hydrograph* can be obtained using a successive number of time steps and computing, with the *runoff equation*, the runoff at the end of each time step from the runoff at the end of the previous time step.

Unit hydrograph The discharge may also be expressed as: $Q = - dS/dT$. Substituting herein the expression of Q in equation (1) gives the differential equation $dS/dT = A.S$, of which the solution is: $S = \exp(-A.t)$. Replacing herein S by Q/A according to equation (1), it is obtained that: $Q = A \exp(-A.t)$. This is called the instantaneous unit hydrograph (IUH) because the Q herein equals Q_2 of the foregoing runoff equation using $R = 0$, and taking S as *unity* which makes Q_1 equal to A according to equation (1). The availability of the foregoing *runoff equation* eliminates the necessity of calculating the *total hydrograph* by the summation of partial hydrographs using the *IUH* as is done with the more complicated convolution method.

Determining the response factor A When the *response factor* A can be determined from the characteristics of the watershed (catchment area), the reservoir can be used as a *deterministic model* or *analytical model*. Otherwise, the factor A can be determined from a data record of rainfall and runoff using the method explained below under *non-linear reservoir*. With this method the reservoir can be used as a black box model.

Conversions 1 mm/day corresponds to 10 m³/day per ha of the watershed 1 l/s per ha corresponds to 8.64 mm/day or 86.4 m³/day per ha

Non-linear Reservoir

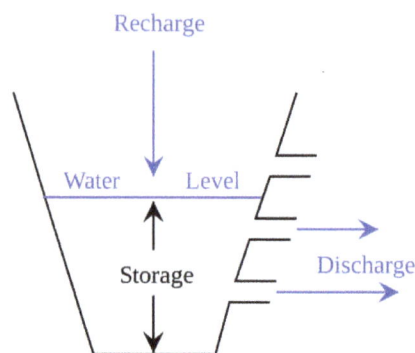

Figure 2. A non-linear reservoir

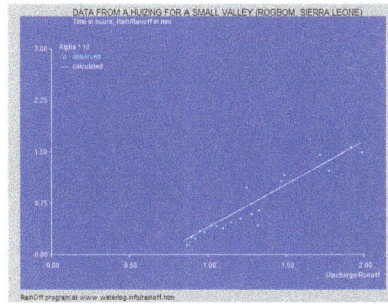

Figure 3. The reaction factor (Aq, Alpha) versus discharge (Q) for a small valley (Rogbom) in Sierra Leone

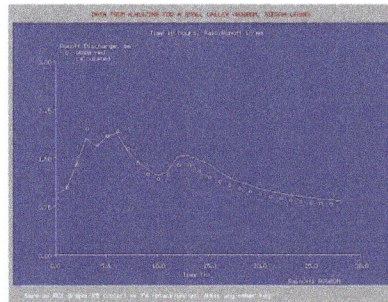

Figure 4. Actual and simulated discharge, Rogbom valley

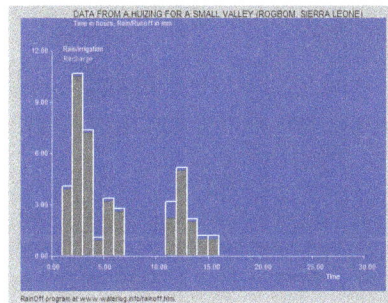

Figure 5. Rainfall and recharge, Rogbom valley

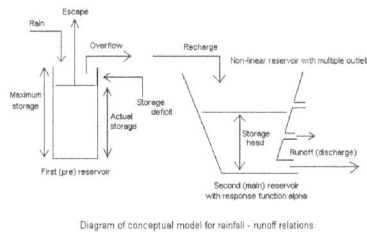

Figure 6. Non-linear reservoir with pre-reservoir for recharge

Contrary to the linear reservoir, the non linear reservoir has a reaction factor A that is not a constant, but it is a function of S or Q (figure 2, 3).

Normally A increases with Q and S because the higher the water level is the higher the discharge capacity becomes. The factor is therefore called Aq instead of A. The non-linear reservoir has *no* usable unit hydrograph.

During periods without rainfall or recharge, i.e. when $R = 0$, the runoff equation reduces to

- $Q_2 = Q_1 \exp \{ - Aq (T_2 - T_1) \}$, or:

or, using a *unit time step* $(T_2 - T_1 = 1)$ and solving for Aq:

- $Aq = - \ln (Q_2/Q_1)$

Hence, the reaction or response factor Aq can be determined from runoff or discharge measurements using *unit time steps* during dry spells, employing a numerical method.

Figure 3 shows the relation between Aq (Alpha) and Q for a small valley (Rogbom) in Sierra Leone. Figure 4 shows observed and *simulated* or *reconstructed* discharge hydrograph of the watercourse at the downstream end of the same valley.

Recharge

The recharge, also called *effective rainfall* or *rainfall excess*, can be modeled by a *pre-reservoir* (figure 6) giving the recharge as *overflow*. The pre-reservoir knows the following elements:

- a maximum storage (Sm) with unit length [L]

- an actual storage (Sa) with unit [L]

- a relative storage: Sr = Sa/Sm

- a maxmimum escape rate (Em) with units length/time [L/T]. It corresponds to the maximum rate of evaporation plus percolation and groundwater recharge, which will not take part in the runoff process (figure 5, 6)

- an actual escape rate: Ea = Sr.Em

- a storage deficiency: Sd = Sm + Ea − Sa

The recharge during a unit time step $(T_2-T_1=1)$ can be found from R = Rain − Sd The actual storage at the end of a *unit time step* is found as Sa2 = Sa1 + Rain − R − Ea, where Sa1 is the actual storage at the start of the time step.

The Curve Number method (CN method) gives another way to calculate the recharge. The *initial abstraction* herein compares with Sm − Si, where Si is the initial value of Sa.

Software

Figures 3 and 4 were made with the RainOff program, designed to analyse rainfall and runoff using the non-linear reservoir model with a pre-reservoir. The program also contains an example of the hydrograph of an agricultural subsurface drainage system for which the value of A can be obtained from the system's characteristics.

V*flo* is another software program for modeling runoff. V*flo* uses radar rainfall and GIS data to generate physics-based, distributed runoff simulation.

The WEAP (Water Evaluation And Planning) software platform models runoff and percolation from climate and land use data, using a choice of linear and non-linear reservoir models.

The RS MINERVE software platform simulates the formation of free surface run-off flow and its propagation in rivers or channels. The software is based on object-oriented programming and allows hydrologic and hydraulic modeling according to a semi-distributed conceptual scheme with different rainfall-runoff model such as HBV, GR4J, SAC-SMA or SOCONT.

Discharge (Hydrology)

In hydrology, discharge is the volume rate of water flow, including any suspended solids (e.g. sediment), dissolved chemicals (e.g. $CaCO_{3(aq)}$), or biologic material (e.g. diatoms), which is transported through a given cross-sectional area.

Synonyms vary by discipline, for example, a fluvial hydrologist studying natural river systems may define discharge as streamflow, whereas an engineer operating a reservoir system might define discharge as outflow, which is contrasted with inflow.

Theory and Calculation

GH Dury and MJ Bradshaw are two hydrologists who devised models showing the relationship between discharge and other variables in a river. The Bradshaw model described how pebble size and other variables change from source to mouth; while Dury considered the relationships between discharge and variables such as slope and friction.

The units that are typically used to express discharge include m³/s (cubic meters per second), ft³/s (cubic feet per second or cfs) and/or acre-feet per day. For example, the average discharge of the Rhine river in Europe is 2,200 cubic metres per second (78,000 cu ft/s) or 190,000,000 cubic metres (150,000 acre·ft) per day.

A commonly applied methodology for measuring, and estimating, the discharge of a river is based on a simplified form of the continuity equation. The equation implies that for any incompressible fluid, such as liquid water, the discharge (Q) is equal to the product of the stream's cross-sectional area (A) and its mean velocity ($\bar{u}$), and is written as:

$$Q = A\bar{u}$$

where

- Q is the discharge ($[LT^{-1}]$; m³/s or ft³/s)

- A is the cross-sectional area of the portion of the channel occupied by the flow ($[L^2]$; m² or ft²)

- $\bar{u}$ is the average flow velocity ($[LT^{-1}]$; m/s or ft/s)

Catchment Discharge

The catchment of a river above a certain location is determined by the surface area of all land which drains toward the river from above that point. The river's discharge at that location depends on the rainfall on the catchment or drainage area and the inflow or outflow of groundwater to or from the area, stream modifications such as dams and irrigation diversions, as well as evaporation and evapotranspiration from the area's land and plant surfaces. In storm hydrology, an important consideration is the stream's discharge hydrograph, a record of how the discharge varies over time after a precipitation event. The stream rises to a peak flow after each precipitation event, then falls in a slow recession. Because the peak flow also corresponds to the maximum water level reached during the event, it is of interest in flood studies. Analysis of the relationship between precipitation intensity and duration and the response of the stream discharge are aided by the concept of the unit hydrograph, which represents the response of stream discharge over time to the application of a hypothetical "unit" amount and duration of rainfall (e.g., half an inch over one hour). The amount of precipitation correlates to the volume of water (depending on the area of the catchment) that subsequently flows out of the river. Using the unit hydrograph method, actual historical rainfalls can be modeled mathematically to confirm characteristics of historical floods, and hypothetical "design storms" can be created for comparison to observed stream responses.

The relationship between the discharge in the stream at a given cross-section and the level of the stream is described by a rating curve. Average velocities and the cross-sectional area of the stream are measured for a given stream level. The velocity and the area give the discharge for that level. After measurements are made for several different levels, a rating table or rating curve may be developed. Once rated, the discharge in the stream may be determined by measuring the level, and determining the corresponding discharge from the rating curve. If a continuous level-recording device is located at a rated cross-section, the stream's discharge may be continuously determined.

Larger flows (higher discharges) can transport more sediment and larger particles downstream than smaller flows due to their greater force. Larger flows can also erode stream banks and damage public infrastructure.

References

- N.A. de Ridder and J. Boonstra, 1994. Analysis of Water Balances. In: H.P.Ritzema (ed.), Drainage Principles and Applications, Publication 16, p.601-634. International Institute for Land Reclamation and Improvement (ILRI), Wageningen, The Netherlands. ISBN 90-70754-33-9

- Should We Build More Large Dams? The Actual Costs of Hydropower Megaproject Development, Energy Policy, March 2014, pp. 1-14

- "External costs of electricity systems (graph format)". ExternE-Pol. Technology Assessment / GaBE (Paul Scherrer Institut). 2005. Archived from the original on 1 November 2013.

- "Upper Siang project likely to be relocated on Chinese concerns". Thehindubusinessline.com. 2006-03-24. Retrieved 2012-07-22.

6

Understanding Hydraulic Engineering

Hydraulic engineering studies the movement of fluids and its application in the design of bridges, water storage facilities, irrigation methods, dams, sewage systems, etc. Hydraulic engineering is involved in predicting the behavior of water and its interaction with these man-made structures. This chapter outlines hydraulic engineering and its importance to the field of civil engineering.

Hydraulic Flood Retention Basin (HFRB)

Hydraulic engineering as a sub-discipline of civil engineering is concerned with the flow and conveyance of fluids, principally water and sewage. One feature of these systems is the extensive use of gravity as the motive force to cause the movement of the fluids. This area of civil engineering is intimately related to the design of bridges, dams, channels, canals, and levees, and to both sanitary and environmental engineering.

Hydraulic engineering is the application of fluid mechanics principles to problems dealing with the collection, storage, control, transport, regulation, measurement, and use of water. Before beginning a hydraulic engineering project, one must figure out how much water is involved. The hydraulic engineer is concerned with the transport of sediment by the river, the interaction of the water with its alluvial boundary, and the occurrence of scour and deposition. "The hydraulic engineer actually develops conceptual designs for the various features which interact with water such as spillways and outlet works for dams, culverts for highways, canals and related structures for irrigation projects, and cooling-water facilities for thermal power plants."

Fundamental Principles

A few examples of the fundamental principles of hydraulic engineering include fluid mechanics, fluid flow, behavior of real fluids, hydrology, pipelines, open channel hydraulics, mechanics of sediment transport, physical modeling, hydraulic machines, and drainage hydraulics.

Fluid Mechanics

Fundamentals of Hydraulic Engineering defines hydrostatics as the study of fluids at rest. In a fluid at rest, there exists a force, known as pressure, that acts upon the fluid's surroundings. This pressure, measured in N/m^2, is not constant throughout the body of fluid. Pressure, p, in a given body of fluid, increases with an increase in depth. Where the upward force on a body acts on the base and can be found by equation:

$$p = \rho gy$$

where,

ρ = density of water

g = specific gravity

y = depth of the body of liquid

Rearranging this equation gives you the pressure head $p/\rho g = y$. Four basic devices for pressure measurement are a piezometer, manometer, differential manometer, Bourdon gauge, as well as an inclined manometer

As Prasuhn states:

> On undisturbed submerged bodies, pressure acts along all surfaces of a body in a liquid, causing equal perpendicular forces in the body to act against the pressure of the liquid. This reaction is known as equilibrium. More advanced applications of pressure are that on plane surfaces, curved surfaces, dams, and quadrant gates, just to name a few.

Behavior of Real Fluids

Real and Ideal Fluids

The main difference between an ideal fluid and a real fluid is that for ideal flow $p_1 = p_2$ and for real flow $p_1 > p_2$. Ideal fluid is incompressible and has no viscosity. Real fluid has viscosity. Ideal fluid is only an imaginary fluid as all fluids that exist have some viscosity.

Viscous Flow

A viscous fluid will deform continuously under a shear force, whereas an ideal fluid doesn't deform.

Laminar Flow and Turbulence

The various effects of disturbance on a viscous flow are stable, transition and unstable.

Bernoulli's Equation

For an ideal fluid, Bernoulli's equation holds along streamlines.

$p/\rho g + u^2/2g = p_1/\rho g + u_1^2/2g = p_2/\rho g + u_2^2/2g$

Boundary Layer

Assuming a flow is bounded on one side only, and that a rectilinear flow passing over a stationary flat plate which lies parallel to the flow, the flow just upstream of the plate has a uniform velocity. As the flow comes into contact with the plate, the layer of fluid actually 'adheres' to a solid surface. There is then a considerable shearing action between the layer of fluid on the plate surface and the second layer of fluid. The second layer is therefore forced to decelerate (though it is not quite brought to rest), creating a shearing action with the third layer of fluid, and so on. As the fluid passes further along the plate, the zone in which shearing action occurs tends to spread further outwards. This zone is known as the 'boundary layer'. The flow outside the boundary layer is free of shear and viscous-related forces so it is assumed to act like an ideal fluid. The intermolecular cohesive forces in a fluid are not great enough to hold fluid together. Hence a fluid will flow under the action of the slightest stress and flow will continue as long as the stress is present. The flow inside the layer can be either viscous or turbulent, depending on Reynolds number.

Applications

Common topics of design for hydraulic engineers include hydraulic structures such as dams, levees, water distribution networks, water collection networks, sewage collection networks, storm water management, sediment transport, and various other topics related to transportation engineering and geotechnical engineering. Equations developed from the principles of fluid dynamics and fluid mechanics are widely utilized by other engineering disciplines such as mechanical, aeronautical and even traffic engineers.

Related branches include hydrology and rheology while related applications include hydraulic modeling, flood mapping, catchment flood management plans, shoreline management plans, estuarine strategies, coastal protection, and flood alleviation.

History

Antiquity

Earliest uses of hydraulic engineering were to irrigate crops and dates back to the Middle East and Africa. Controlling the movement and supply of water for growing food has been used for many thousands of years. One of the earliest hydraulic machines, the water clock was used in the early 2nd millennium BC. Other early examples of using gravity to move water include the Qanat system in ancient Persia and the very similar Turpan water system in ancient China as well as irrigation canals in Peru.

In ancient China, hydraulic engineering was highly developed, and engineers constructed massive canals with levees and dams to channel the flow of water for irrigation, as well as locks to allow ships to pass through. Sunshu Ao is considered the first Chinese hydraulic engineer. Another important Hydraulic Engineer in China, Ximen Bao was credited of starting the practice of large scale canal irrigation during the Warring States period (481 BC-221 BC), even today hydraulic engineers remain a respectable position in China. Before becoming General Secretary of the Communist Party of China in 2002, Hu Jintao was a hydraulic engineer and holds an engineering degree from Tsinghua University

Eupalinos of Megara, was an ancient Greek engineer who built the Tunnel of Eupalinos on Samos in the 6th century BC, an important feat of both civil and hydraulic engineering. The civil engineering aspect of this tunnel was the fact that it was dug from both ends which required the diggers to maintain an accurate path so that the two tunnels met and that the entire effort maintained a sufficient slope to allow the water to flow.

Hydraulic engineering was highly developed in Europe under the aegis of the Roman Empire where it was especially applied to the construction and maintenance of aqueducts to supply water to and remove sewage from their cities. In addition to supplying the needs of their citizens they used hydraulic mining methods to prospect and extract alluvial gold deposits in a technique known as hushing, and applied the methods to other ores such as those of tin and lead.

In the 15th century, the Somali Ajuran Empire was the only hydraulic empire in Africa. As a hydraulic empire, the Ajuran State monopolized the water resources of the Jubba and Shebelle Rivers. Through hydraulic engineering, it also constructed many of the limestone wells and cisterns of the state that are still operative and in use today. The rulers developed new systems for agriculture and taxation, which continued to be used in parts of the Horn of Africa as late as the 19th century.

Further advances in hydraulic engineering occurred in the Muslim world between the 8th to 16th centuries, during what is known as the Islamic Golden Age. Of particular importance was the 'water management technological complex' which was central to the Islamic Green Revolution and, by extension, a precondition for the emergence of modern technology. The various components of this 'toolkit' were developed in different parts of the Afro-Eurasian landmass, both within and beyond the Islamic world. However, it was in the medieval Islamic lands where the technological complex was assembled and standardized, and subsequently diffused to the rest of the Old World. Under the rule of a single Islamic Caliphate, different regional hydraulic technologies were assembled into "an identifiable water management technological complex that was to have a global impact." The various components of this complex included canals, dams, the *qanat* system from Persia, regional water-lifting devices such as the *noria, shaduf* and screwpump from Egypt, and the windmill from Islamic Afghanistan. Other original Islamic developments included the *saqiya* with a flywheel effect from Islamic Spain, the reciprocating suction pump and crankshaft-connecting rod mechanism from Iraq, the geared and hydropowered water supply system from Syria, and the water purification methods of Islamic chemists.

Modern Times

In many respects the fundamentals of hydraulic engineering haven't changed since ancient times. Liquids are still moved for the most part by gravity through systems of canals and aqueducts, though the supply reservoirs may now be filled using pumps. The need for water has steadily increased from ancient times and the role of the hydraulic engineer is a critical one in supplying it. For example, without the efforts of people like William Mulholland the Los Angeles area would not have been able to grow as it has because it simply doesn't have enough local water to support its population. The same is true for many of our world's largest cities. In much the same way, the central valley of California could not have become such an important agricultural region without effective water management and distribution for irrigation. In a somewhat parallel way to what happened in California the creation of the Tennessee Valley Authority(TVA) brought work and prosperity to the South by building dams to generate cheap electricity and control flooding in the region, making rivers navigable and generally modernizing life in the region.

Leonardo da Vinci (1452–1519) performed experiments, investigated and speculated on waves and jets, eddies and streamlining. Isaac Newton (1642–1727) by formulating the laws of motion and his law of viscosity, in addition to developing the calculus, paved the way for many great developments in fluid mechanics. Using Newton's laws of motion, numerous 18th-century mathematicians solved many frictionless (zero-viscosity) flow problems. However, most flows are dominated by viscous effects, so engineers of the 17th and 18th centuries found the inviscid flow solutions unsuitable, and by experimentation they developed empirical equations, thus establishing the science of hydraulics.

Late in the 19th century, the importance of dimensionless numbers and their relationship to turbulence was recognized, and dimensional analysis was born. In 1904 Ludwig Prandtl published a key paper, proposing that the flow fields of low-viscosity fluids be divided into two zones, namely a thin, viscosity-dominated boundary layer near solid surfaces, and an effectively inviscid outer zone away from the boundaries. This concept explained many former paradoxes, and enabled subsequent engineers to analyze far more complex flows. However, we still have no complete theory for the nature of turbulence, and so modern fluid mechanics continues to be combination of experimental results and theory.

The modern hydraulic engineer uses the same kinds of computer-aided design (CAD) tools as many of the other engineering disciplines while also making use of technologies like computational fluid dynamics to perform the calculations to accurately predict flow characteristics, GPS mapping to assist in locating the best paths for installing a system and laser-based surveying tools to aid in the actual construction of a system.

References

- Sally Ganchy, Sarah Gancher (2009), Islam and Science, Medicine, and Technology, The Rosen Publishing Group, p. 41, ISBN 1-4358-5066-1

- Howard R. Turner (1997), Science in Medieval Islam: An Illustrated Introduction, p. 181, University of Texas Press, ISBN 0-292-78149-0

7

Applications of Hydraulic Engineering

Hydraulic engineering is principally employed in the construction of storm drains, levees and dams. It is used to study phenomenon like sediment transport and applying it to civil engineering structures related to water. This chapter also explores the uses and principles of the rain gauge, stream gauge, hygrometer and Green Kenue. The chapter strategically encompasses and incorporates the major components and key concepts of hydraulic engineering, providing a complete understanding.

Storm Drain

Storm drain with its pipe visible beneath it due to construction work

A storm drain, storm sewer (US), surface water drain/sewer (UK), or stormwater drain (Australia and New Zealand) is designed to drain excess rain and ground water from impervious surfaces such as paved streets, car parks, parking lots, footpaths, sidewalks, and roofs. Storm drains vary in design from small residential dry wells to large municipal systems. They are fed by street gutters on most motorways, freeways and other busy roads, as well as towns in areas which experience heavy rainfall, flooding and coastal towns which experience regular storms. Even the gutters from houses and buildings can be connected to the Storm drain. Many storm drainage systems are designed to drain the storm water, untreated, into rivers or streams. As a result, it is not acceptable to pour certain types of chemicals into the drains.

Some storm drains lead to a mixing of stormwater (rainwater) with sewage, either intentionally – in the case of combined sewers – or unintentionally.

Nomenclature

There are several related terms which are used differently in American and British English:

	American English	**British English**	**Comments**
Combined sewer	A sewer designed and intended to serve as a sanitary sewer and a storm sewer, or as an industrial sewer and a storm sewer.	Same as American English	Stormwater mixed with sewage
Storm sewer	A sewer designed and intended to carry only storm waters, surface run-off, street wash waters, and drainage.	See surface water sewer or surface sewer.	Only storm-water
Surface water sewer or surface sewer	See storm sewer	A sewer designed and intended to carry only rainwater runoff.	Only storm-water
Stormwater bypass	Same as British English	A combined sewer discharge pipe-line intended to bypass wastewater treatment plants during a peak runoff events.	Stormwater mixed with sewage
Road channel	See roadside ditch	A roadside channel to prevent uncontrolled runoff along roadway surfaces.	Only storm-water
Road gully	See roadside ditch	Consists of a gully grating on a chamber that connects to a surface water sewer / drain, ditch, or watercourse	Only storm-water
Roadside ditch	A roadside channel to prevent uncontrolled runoff along roadway surfaces.	See road gully	Only storm-water

Function

American-style curbside storm drain receiving urban runoff

Inlet

Full view of a storm drain (Ontario, Canada)

There are two main types of stormwater drain (highway drain or road gully in the UK) inlets: side inlets and grated inlets. Side inlets are located adjacent to the curb (kerb) and rely on the ability of the opening under the backstone or lintel to capture flow. They are usually depressed at the invert of the channel to improve capture capacity.

Many inlets have gratings or grids to prevent people, vehicles, large objects or debris from falling into the storm drain. Grate bars are spaced so that the flow of water is not impeded, but sediment and many small objects can also fall through. However, if grate bars are too far apart, the openings may present a risk to pedestrians, bicyclists, and others in the vicinity. Grates with long narrow slots parallel to traffic flow are of particular concern to cyclists, as the front tire of a bicycle may become stuck, causing the cyclist to go over the handlebars or lose control and fall. Storm drains in streets and parking areas must be strong enough to support the weight of vehicles, and are often made of cast iron or reinforced concrete.

Old German storm drain in Küstrin (now Kostrzyn nad Odrą in Poland)

Some of the heavier sediment and small objects may settle in a catchbasin, or sump, which lies immediately below the outlet, where water from the top of the catchbasin reservoir overflows into the sewer proper. The catchbasin serves much the same function as the "trap" in household wastewater plumbing in trapping objects.

In the United States, unlike the plumbing trap, the catchbasin does not necessarily prevent sewer gases such as hydrogen sulfide and methane from escaping. However, in the United Kingdom, where they are called gully pots, they are designed as true water-filled traps and do block the egress of gases and rodents.

Most catchbasins will contain stagnant water during the drier parts of the year, and can, in warm countries, be used by mosquitos for breeding. Larvicides or disruptive larval hormones, sometimes released from "mosquito biscuits", have been used to control mosquito breeding in catchbasins. Mosquitoes may be physically prevented from reaching the standing water or migrating into the sewer proper by the use of an "inverted cone filter". Another method of mosquito control is to spread a thin layer of oil on the surface of stagnant water, interfering with the breathing tubes of mosquito larvae.

The performance of catchbasins at removing sediment and other pollutants depends on the design of the catchbasin (for example, the size of the sump), and on routine maintenance to retain the storage available in the sump to capture sediment. Municipalities typically have large vacuum trucks that perform this task.

Catchbasins act as a first-line pretreatment for other treatment practices, such as retention basins, by capturing large sediments and street litter from urban runoff before it enters the storm drainage pipes.

A storm sewer under the main road empties into a bigger open channel

Piping

Pipes can come in many different cross-sectional shapes (rectangular, square, bread-loaf-shaped, oval, inverted pear-shaped, egg shaped, and most commonly, circular). Drainage systems may have many different features including waterfalls, stairways, balconies and pits for catching rubbish, sometimes called Gross Pollutant Traps (GPTs). Pipes made of different materials can also be used, such as brick, concrete, high-density polyethylene or galvanized steel. Fibre reinforced plastic is being used more commonly for drain pipes and fittings.

Outlet

Most drains have a single large exit at their point of discharge (often covered by a grating) into a canal, river, lake, reservoir, sea or ocean. Other than catchbasins, typically there are no treatment facilities in the piping system. Small storm drains may discharge into individual dry wells. Storm drains may be interconnected using slotted pipe, to make a larger dry well system. Storm drains may discharge into man-made excavations known as recharge basins or retention ponds.

Environmental Impacts

Water Quantity

Storm drains are often unable to manage the quantity of rain that falls during heavy rains and/or storms. When storm drains are inundated, basement and street flooding can occur. Unlike catastrophic flooding events, this type of urban flooding occurs in built-up areas where man-made drainage systems are prevalent. Urban flooding is the primary cause of sewer backups and basement flooding which can affect properties year after year.

The *first flush* from urban runoff can be extremely dirty. Storm water may become contaminated while running down the road or other impervious surface, or from lawn chemical run-off, before entering the drain.

Water running off these impervious surfaces tends to pick up gasoline, motor oil, heavy metals, trash and other pollutants from roadways and parking lots, as well as fertilizers and pesticides from lawns. Roads and parking lots are major sources of nickel, copper, zinc, cadmium, lead and polycyclic aromatic hydrocarbons (PAHs), which are created as combustion byproducts of gasoline and other fossil fuels. Roof runoff contributes high levels of synthetic organic compounds and zinc (from galvanized gutters). Fertilizer use on residential lawns, parks and golf courses is a significant source of nitrates and phosphorus.

Separation of undesired runoff can be achieved by installing devices within the storm sewer system. These devices are relatively new and can only be installed with new development or during major upgrades. They are referred to as oil-grit separators (OGS) or oil-sediment separators (OSS). They consist of a specialized manhole chamber, and use the water flow and/or gravity to separate oil and grit.

Reducing Stormwater Flows

Runoff into storm sewers can be minimized by including *sustainable urban drainage systems* (UK term) or *low impact development* or *green infrastructure* practices (US terms) into municipal plans. To reduce stormwater from rooftops, flows from eaves troughs (rain gutters and downspouts) may be infiltrated into adjacent soil, rather than discharged into the storm sewer system. Storm water runoff from paved surfaces can be directed to unlined ditches (sometimes called swales or bioswales) before flowing into the storm sewers, again to allow the runoff to soak into the ground. Permeable paving materials can be used in building sidewalks, driveways and in some cases, parking lots, to infiltrate a portion of the stormwater volume.

In many areas, detention tanks are required to be installed inside a property and are used to temporarily hold rainwater runoff during heavy rains and restrict the outlet flow to the public sewer. This lessens the risk of the public sewer being overburdened during heavy rain. An overflow outlet may also be utilized which connects higher on the outlet side of the detention tank. This overflow would prevent the detention tank from completely filling up. By restricting the flow of water in this way and temporarily holding the water in a detention tank public sewers are far less likely to become surcharged.

Mosquito Breeding

Catch basins are commonly designed with a sump area below the outlet pipe level which is a reservoir for water and debris to help prevent the pipe from clogging. Unless they are constructed with permeable bottoms to allow water to infiltrate into the underlying soil, this subterranean basin can become a perfect mosquito breeding area because it is cool, dark, and retains stagnant water for long periods of time. Combined with standard grates which have holes large enough for mosquitoes to enter and leave the basin, this is a major problem in mosquito control.

Basins can be filled with concrete up to the pipe level to prevent this reservoir from forming. Without proper maintenance, the functionality of the basin is questionable, as these catch basins are most commonly not cleaned annually as is needed to make them perform as designed. The trapping of debris serves no purpose because once filled they operate as if no basins were present, but continue to allow a shallow area of water retention for the breeding of mosquito. Moreover, even if cleaned and maintained, the water reservoir remains filled, accommodating the breeding of mosquitoes.

Relationship to Sanitary Sewer Systems

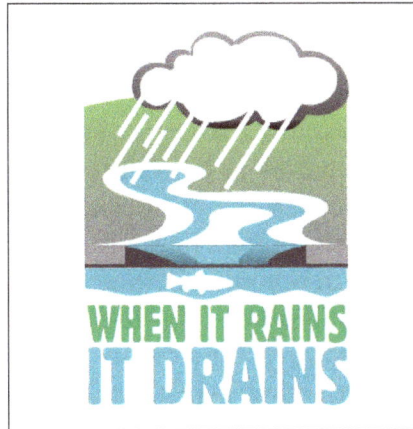

Sign alerting public to avoid dumping waste into storm drains

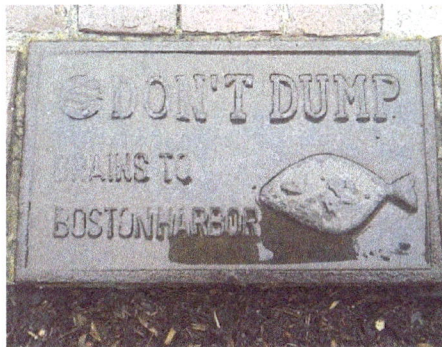

Typical signage embedded in pavement next to a storm drain in Boston, in the United States

Storm drains are separate and distinct from sanitary sewer systems. The separation of storm sewers from sanitary sewers helps to prevent sewage treatment plants becoming overwhelmed by infiltration/inflow during a rainstorm, which can result in untreated sewage being discharged into the environment.

Many storm drainage systems are designed to drain the storm water, untreated, into rivers or streams. Many local governments conduct public awareness campaigns about this, lest waste be dumped into the storm drain system. In the city of Cleveland, Ohio, for example, all new catch basins installed have inscriptions on them not to dump any waste, and usually include a fish imprint as well. Trout Unlimited Canada recommends that a yellow fish symbol be painted next to existing storm drains.

Combined Sewers

Cities that installed their sewage collection systems before the 1930s typically used single piping systems to transport both urban runoff and sewage. This type of collection system is referred to as a combined sewer system (CSS). The cities' rationale when combined sewers were built was that it would be cheaper to build just a single system. In these systems a sudden large rainfall that exceeds sewage treatment capacity will be allowed to overflow directly from the storm drains into receiving waters via structures called combined sewer overflows.

Storm drains are typically at shallower depths than combined sewers; because, while storm drains are designed to accept surface runoff from streets, combined sewers were designed to also accept sewage flows from buildings with basements.

New York City, Washington DC, Seattle and other cities with combined systems have this problem due to a large influx of storm water after every heavy rain. Some cities have dealt with this by adding large storage tanks or ponds to hold the water until it can be treated. Chicago has a system of tunnels, collectively called the Deep Tunnel, underneath the city for storing its stormwater. In many areas detention tanks or roof detention systems are required to be installed for a property and are used to temporarily hold rainwater runoff during heavy rains and restrict the outlet flow to the public sewer. This lessens the risk of the public sewer being overburdened during a heavy rain. An overflow outlet may also be utilized which connects higher on the outlet side of the detention tank. This overflow would prevent the detention tank from completely filling up. By restricting the flow of water in this way and temporarily holding the water in a detention tank or by roof detention public sewers are far less likely to become surcharged.

Regulations and Local Building Codes

Building codes and local government ordinances vary greatly on the handling of storm drain run-off. New developments might be required to construct their own storm drain processing capacity for returning the runoff to the water table and bioswales may be required in sensitive ecological areas to protect the watershed.

In the United States, cities, suburban communities and towns with over 10,000 population are required to obtain discharge permits for their storm sewer systems, under the Clean Water Act. The Environmental Protection Agency (EPA) issued stormwater regulations for large cities in 1990 and for other communities in 1999. The permits require local governments to operate stormwater management programs, covering both construction of new buildings and facilities, and maintenance of their existing municipal drainage networks. Many municipalities have revised their local ordinances covering management of runoff. State government facilities, such as roads and highways, are also subject to the stormwater management regulations. Many local municipalities have placed ordinances for both commercial and residential stormwater management practices to be designed, implemented, and approved before an occupancy permit is released.

Examples

Southeastern Los Angeles County installed thousands of stainless steel, full-capture trash devices on their road drains in 2011.

Exploration

An international subculture has grown up around the exploration of stormwater drains. Societies such as the Cave Clan regularly explore the drains underneath cities. This is commonly known as "urban exploration," but is also known as *draining* when in specific relation to storm drains.

Residence

In several large American cities, homeless people live in storm drains. At least 300 people live in the 200 miles of underground storm drains of Las Vegas, many of them making a living finding unclaimed winnings in the gambling machines. An organization called Shine a Light was founded in 2009 to help the drain residents after over 20 drowning deaths occurred in the preceding years. A man in San Diego was evicted from a storm drain after living there for nine months in 1986.

History

Ancient Roman gully hole in Ostia Antica in Italy

Archaeological studies have revealed use of rather sophisticated stormwater runoff systems in ancient cultures. For example, in Minoan Crete approximately 4000 years before present, cities such as Phaistos were designed to have storm drains and channels to collect precipitation runoff. At Cretan Knossos, storm drains include stone-lined structures large enough for a person to crawl through. Other examples of early civilizations with elements of stormwater drain systems include early people of Mainland Orkney such as Gurness and the Brough of Birsay in Scotland.

Levee

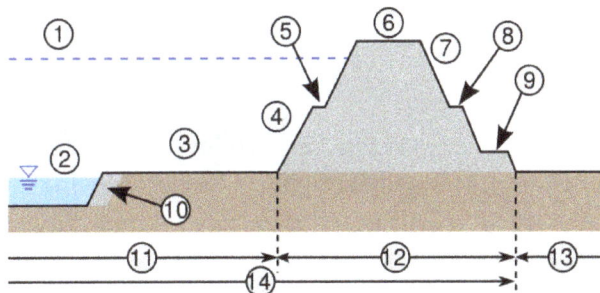

1. Design High Water Level (HWL) 2. Low water channel 3. Flood channel
4. Riverside Slope 5. Riverside Banquette 6. Levee Crown 7. Landside Slope
8. Landside Banquette 9. Berm 10. Low water revetment 11. Riverside land
12. Levee 13. Protected lowland 14. River zone

The side of a levee in Sacramento, California

A levee, levée, dike, dyke, embankment, floodbank or stopbank is an elongated naturally occurring ridge or artificially constructed fill or wall, which regulates water levels. It is usually earthen and often parallel to the course of a river in its floodplain or along low-lying coastlines.

Etymology

Levee

The word *levee*, from the French word *levée* (from the feminine past participle of the French verb *lever*, "to raise"), is used in American English (notably in the Midwest and Deep South). It originated in New Orleans a few years after the city's founding in 1718 and was later adopted by English speakers. The name derives from the trait of the levee's ridges being *raised* higher than both the channel and the surrounding floodplains.

Dike

The modern word *dike* or *dyke* most likely derives from the Dutch word "*dijk*", with the construction of dikes in Frisia (now part of the Netherlands and Germany) well attested as early as the 11th century. The 126 kilometres (78 mi) long Westfriese Omringdijk was completed by 1250, and was formed by connecting existing older dikes. The Roman chronicler Tacitus even mentions that the rebellious Batavi pierced dikes to flood their land and to protect their retreat (AD 70). The word *dijk* originally indicated both the trench and the bank. It is closely related to the English verb *to dig*.

In Anglo-Saxon, the word *dic* already existed and was pronounced as *dick* in northern England and as *ditch* in the south. Similar to Dutch, the English origins of the word lie in digging a trench and forming the upcast soil into a bank alongside it. This practice has meant that the name may be given to either the excavation or the bank. Thus Offa's Dyke is a combined structure and Car Dyke is a trench though it once had raised banks as well. In the midlands and north of England, and in the United States, a dike is what a ditch is in the south, a property boundary marker or small drainage channel. Where it carries a stream, it may be called a running dike as in *Rippingale Running Dike*, which leads water from the catchwater drain, Car Dyke, to the South Forty Foot Drain in Lincolnshire (TF1427). The Weir Dike is a soak dike in Bourne North Fen, near Twenty and alongside the River Glen, Lincolnshire. In the Norfolk and Suffolk Broads, a dyke may be a drainage ditch or a narrow artificial channel off a river or broad for access or mooring, some longer dykes being named, e.g. Candle Dyke.

In parts of Britain, particularly Scotland, a dyke may be a field wall, generally made with dry stone.

Usage

A reinforced embankment

The main purpose of artificial levees is to prevent flooding of the adjoining countryside and to slow natural course changes in a waterway to provide reliable shipping lanes for maritime commerce over time; they also confine the flow of the river, resulting in higher and faster water flow. Levees can be mainly found along the sea, where dunes are not strong enough, along rivers for protection against high-floods, along lakes or along polders. Furthermore, levees have been built for the purpose of empoldering, or as a boundary for an inundation area. The latter can be a controlled inundation by the military or a measure to prevent inundation of a larger area surrounded by levees. Levees have also been built as field boundaries and as military defences.

Levees can be permanent earthworks or emergency constructions (often of sandbags) built hastily in a flood emergency. When such an emergency bank is added on top of an existing levee it is known as a *cradge*.

Some of the earliest levees were constructed by the Indus Valley Civilization (in Pakistan and North India from circa 2600 BC) on which the agrarian life of the Harappan peoples depended. Levees were also constructed over 3,000 years ago in ancient Egypt, where a system of levees was built along the left bank of the River Nile for more than 600 miles (970 km), stretching from modern Aswan to the Nile Delta on the shores of the Mediterranean. The Mesopotamian civilizations and ancient China also built large levee systems. Because a levee is only as strong as its weakest point, the height and standards of construction have to be consistent along its length. Some authorities have argued that this requires a strong governing authority to guide the work, and may have been a catalyst for the development of systems of governance in early civilizations. However, others point to evidence of large scale water-control earthen works such as canals and/or levees dating from before King Scorpion in Predynastic Egypt, during which governance was far less centralized.

Levees are usually built by piling earth on a cleared, level surface. Broad at the base, they taper to a level top, where temporary embankments or sandbags can be placed. Because flood discharge intensity increases in levees on both river banks, and because silt deposits raise the level of river-beds, planning and auxiliary measures are vital. Sections are often set back from the river to form a wider channel, and flood valley basins are divided by multiple levees to prevent a single breach from flooding a large area. A levee made from stones laid in horizontal rows with a bed of thin turf between each of them is known as a *spetchel*.

Artificial levees require substantial engineering. Their surface must be protected from erosion, so they are planted with vegetation such as Bermuda grass in order to bind the earth together. On the land side of high levees, a low terrace of earth known as a *banquette* is usually added as another anti-erosion measure. On the river side, erosion from strong waves or currents presents an even greater threat to the integrity of the levee. The effects of erosion are countered by planting suitable vegetation or installing stones, boulders, weighted matting or concrete revetments. Separate ditches or drainage tiles are constructed to ensure that the foundation does not become waterlogged.

River Flood Prevention

Broken levee on the Sacramento River

A levee keeps high water on the Mississippi River from flooding Gretna, Louisiana, in March 2005.

Prominent levee systems have been built along the Mississippi River and Sacramento River in the United States, and the Po, Rhine, Meuse River, Rhone, Loire, Vistula, the delta formed by the Rhine, Maas/Meuse and Scheldt in the Netherlands and the Danube in Europe.

The Mississippi levee system represents one of the largest such systems found anywhere in the world. It comprises over 3,500 miles (5,600 km) of levees extending some 1,000 kilometres (620 mi) along the Mississippi, stretching from Cape Girardeau, Missouri, to the Mississippi Delta. They were begun by French settlers in Louisiana in the 18th century to protect the city of New Orleans. The first Louisiana levees were about 3 feet (0.91 m) high and covered a distance of about 50 miles (80 km) along the riverside. The U.S. Army Corps of Engineers, in conjunction with the Mississippi River Commission, extended the levee system beginning in 1882 to cover the riverbanks from Cairo, Illinois to the mouth of the Mississippi delta in Louisiana. By the mid-1980s, they had reached their present extent and averaged 24 feet (7.3 m) in height; some Mississippi levees are as high as 50 feet (15 m). The Mississippi levees also include some of the longest continuous individual levees in the world. One such levee extends southwards from Pine Bluff, Arkansas, for a distance of some 380 miles (610 km).

Soil Reinforcement and Levee Protection – The United States Army Corps of Engineers (USACE) recommends and supports Cellular Confinement technology (geocells) as a best management practice. Particular attention is given to the matter of surface erosion, overtopping prevention and protection of levee crest and downstream slope. Reinforcement with geocells provides tensile force to the soil to better resist instability.

Effects of Levees Upon the Elevation of the River Bed

Artificial levees can lead to an elevation of the natural river bed over time; whether this happens or not and how fast, depends on different factors, one of them being the amount and type of the bed load of a river. Alluvial rivers with intense accumulations of sediment tend to this behavior. Examples of rivers where artificial levees led to an elevation of the river bed, even up to a point where the river bed is higher than the adjacent ground surface behind the levees, are found for the Yellow River in China and the Mississippi in the USA.

Coastal Flood Prevention

Levees are very common on the marshlands bordering the Bay of Fundy in New Brunswick and Nova Scotia Canada. The Acadians who settled the area can be credited with the original construction of many of the levees in the area, created for the purpose of farming the fertile tidal marshlands. These levees are referred to as dykes. They are constructed with hinged sluice gates that open on the falling tide to drain freshwater from the agricultural marshlands, and close on the rising tide to prevent seawater from entering behind the dyke. These sluice gates are called "aboiteaux". In the Lower Mainland around the city of Vancouver, British Columbia, there are levees (known locally as dikes, and also referred to as "the sea wall") to protect low-lying land in the Fraser River delta, particularly the city of Richmond on Lulu Island. There are also dikes to protect other locations which have flooded in the past, such as the Pitt Polder, land adjacent to the Pitt River and other tributary rivers.

Coastal flood prevention levees are also common along the inland coastline behind the Wadden Sea, an area devastated by many historic floods. Thus the peoples and governments have erected increasingly large and complex flood protection levee systems to stop the sea even during storm floods. The biggest of these are of course the huge levees in the Netherlands, which have gone beyond just defending against floods, as they have aggressively taken back land that is below mean sea level.

Spur Dykes or Groynes

These typically man-made hydraulic structures are situated to protect against erosion. They are typically placed in alluvial rivers perpendicular, or at an angle, to the bank of the channel or the revetment, and are used widely along coastlines. There are two common types of spur dyke, permeable and impermeable, depending on the materials used to construct them.

Natural Levees

Natural levees commonly form around lowland rivers and creeks without human intervention. They are elongate ridges of mud and/or silt that form on the river floodplains immediately adjacent to the cut banks. Like artificial levees, they act to reduce the likelihood of floodplain inundation.

Deposition of levees is a natural consequence of the flooding of meandering rivers which carry high proportions of suspended sediment in the form of fine sands, silts, and muds. Because the carrying capacity of a river depends in part on its depth, the sediment in the water which is over the flooded banks of the channel is no longer capable of keeping the same amount of fine sediments in suspension as the main thalweg. The extra fine sediments thus settle out quickly on the parts of the floodplain nearest to the channel. Over a significant number of floods, this will eventually result in the building up of ridges in these positions, and reducing the likelihood of further floods and episodes of levee building.

If aggradation continues to occur in the main channel, this will make levee overtopping more likely again, and the levees can continue to build up. In some cases this can result in the channel bed eventually rising above the surrounding floodplains, penned in only by the levees around it; an example is the Yellow River in China near the sea, where oceangoing ships appear to sail high above the plain on the elevated river.

Levees are common in any river with a high suspended sediment fraction, and thus are intimately associated with meandering channels, which also are more likely to occur where a river carries large fractions of suspended sediment. For similar reasons, they are also common in tidal creeks, where tides bring in large amounts of coastal silts and muds. High spring tides will cause flooding, and result in the building up of levees.

Levee Failures and Breaches

Both natural and man-made levees can fail in a number of ways. Factors that cause levee failure include overtopping, erosion, structural failures, and levee saturation. The most frequent (and dangerous) is a *levee breach*. Here, a part of the levee actually breaks or is eroded away, leaving a large opening for water to flood land otherwise protected by the levee. A breach can be a sudden or gradual failure, caused either by surface erosion or by subsurface weakness in the levee. A breach can leave a fan-shaped deposit of sediment radiating away from the breach, described as a crevasse splay. In natural levees, once a breach has occurred, the gap in the levee will remain until it is again filled in by levee building processes. This increases the chances of future breaches occurring in the same location. Breaches can be the location of meander cutoffs if the river flow direction is permanently diverted through the gap.

Sometimes levees are said to fail when water *overtops* the crest of the levee. This will cause flooding on the floodplains, but because it does not damage the levee, it has fewer consequences for future flooding.

Among various failure mechanisms that cause levee breaches, soil erosion is found to be one of the most important factors . Predicting soil erosion and scour generation when overtopping happens is important in order to design stable levee and floodwalls. There have been numerous studies to investigate the erodibility of soils. Briaud et al. (2008) used Erosion Function Apparatus (EFA) test to measure the erodibility of the soils and afterwards by using Chen 3D software, numerical simulations were performed on the levee to find out the velocity vectors in the overtopping water and the generated scour when the overtopping water impinges the levee. By analyzing the results from EFA test, an erosion chart to categorize erodibility of the soils was developed. Hughes and Nadal in 2009 studied the effect of combination of wave overtopping and storm surge overflow on the erosion and scour generation in levees. The study included hydraulic parameters and flow characteristics such as flow thickness, wave intervals, surge level above levee crown in analyzing scour development. According to the laboratory tests, empirical correlations related to average overtopping discharge were derived to analyze the resistance of levee against erosion. These equations could only fit to the situation similar to the experimental tests while they can give a reasonable estimation if applied to other conditions. Osouli et al. (2014) and Karimpour et al. (2015) conducted lab scale physical modeling of levees to evaluate score characterization of different levees due to floodwall overtopping.

Dam

A dam is a barrier that impounds water or underground streams. Reservoirs created by dams not only suppress floods but also provide water for activities such as irrigation, human consumption, industrial use, aquaculture, and navigability. Hydropower is often used in conjunction with dams to generate electricity. A dam can also be used to collect water or for storage of water which can be evenly distributed between locations. Dams generally serve the primary purpose of retaining water, while other structures such as floodgates or levees (also known as dikes) are used to manage or prevent water flow into specific land regions.

Glen Canyon Dam

A sideview of the Lake Vyrnwy dam, in Wales, finished in 1888

The word *dam* can be traced back to Middle English, and before that, from Middle Dutch, as seen in the names of many old cities.

History

Ancient Dams

Early dam building took place in Mesopotamia and the Middle East. Dams were used to control the water level, for Mesopotamia's weather affected the Tigris and Euphrates rivers.

The earliest known dam is the Jawa Dam in Jordan, 100 kilometres (62 mi) northeast of the capital Amman. This gravity dam featured an originally 9-metre-high (30 ft) and 1 m-wide (3.3 ft) stone wall, supported by a 50 m-wide (160 ft) earth rampart. The structure is dated to 3000 BC.

The Ancient Egyptian Sadd-el-Kafara Dam at Wadi Al-Garawi, located about 25 km (16 mi) south of Cairo, was 102 m (335 ft) long at its base and 87 m (285 ft) wide. The structure was built around 2800 or 2600 BC as a diversion dam for flood control, but was destroyed by heavy rain during construction or shortly afterwards. During the XIIth dynasty in the 19th century BC, the Pharaohs Senosert III, Amenemhat III and Amenemhat IV dug a canal 16 km (9.9 mi) long linking the Fayum Depression to the Nile in Middle Egypt. Two dams called Ha-Uar running east-west were built to retain water during the annual flood and then release it to surrounding lands. The lake called "Mer-wer" or Lake Moeris covered 1,700 km² (660 sq mi) and is known today as Berkat Qaroun.

One of the engineering wonders of the ancient world was the Great Dam of Marib in Yemen. Initiated somewhere between 1750 and 1700 BC, it was made of packed earth - triangular in cross section, 580 m (1,900 ft) in length and originally 4 m (13 ft) high - running between two groups of rocks on either side, to which it was linked by substantial stonework. Repairs were carried out during various periods, most important around 750 BC, and 250 years later the dam height was increased to 7 m (23 ft). After the end of the Kingdom of Saba, the dam fell under the control of the Ḥimyarites (~115 BC) who undertook further improvements, creating a structure 14 m (46 ft) high, with five spillway channels, two masonry-reinforced sluices, a settling pond, and a 1,000 m (3,300 ft) canal to a distribution tank. These extensive works were not actually finalized until 325 AD and allowed the irrigation of 25,000 acres (100 km²).

By the mid-late 3rd century BC, an intricate water-management system within Dholavira in modern-day India was built. The system included 16 reservoirs, dams and various channels for collecting water and storing it.

Eflatun Pınar is a Hittite dam and spring temple near Konya, Turkey. It is thought to be from the time of the Hittite empire between the 15th and 13th century BC.

The Kallanai is constructed of unhewn stone, over 300 m (980 ft) long, 4.5 m (15 ft) high and 20 m (66 ft) wide, across the main stream of the Kaveri river in Tamil Nadu, South India. The basic structure dates to the 2nd century AD and is considered one of the oldest water-diversion or water-regulator structures in the world which is still in use. The purpose of the dam was to divert the waters of the Kaveri across the fertile delta region for irrigation via canals.

Du Jiang Yan is the oldest surviving irrigation system in China that included a dam that directed waterflow. It was finished in 251 BC. A large earthen dam, made by Sunshu Ao, the prime minister of Chu (state), flooded a valley in modern-day northern Anhui province that created an enormous irrigation reservoir (100 km (62 mi) in circumference), a reservoir that is still present today.

Roman Engineering

The Roman dam at Cornalvo in Spain has been in use for almost two millennia.

Roman dam construction was characterized by "the Romans' ability to plan and organize engineering construction on a grand scale." Roman planners introduced the then-novel concept of large reservoir dams which could secure a permanent water supply for urban settlements over the dry season. Their pioneering use of water-proof hydraulic mortar and particularly Roman concrete allowed for much larger dam structures than previously built, such as the Lake Homs Dam, possibly the largest water barrier to that date, and the Harbaqa Dam, both in Roman Syria. The highest Roman dam was the Subiaco Dam near Rome; its record height of 50 m (160 ft) remained unsurpassed until its accidental destruction in 1305.

Roman engineers made routine use of ancient standard designs like embankment dams and masonry gravity dams. Apart from that, they displayed a high degree of inventiveness, introducing most of the other basic dam designs which had been unknown until then. These include arch-gravity dams, arch dams, buttress dams and multiple arch buttress dams, all of which were known and employed by the 2nd century AD. Roman workforces also were the first to build dam bridges, such as the Bridge of Valerian in Iran.

Remains of the Band-e Kaisar dam, built by the Romans in the 3rd century AD

In Iran, bridge dams such as the Band-e Kaisar were used to provide hydropower through water wheels, which often powered water-raising mechanisms. One of the first was the Roman-built dam bridge in Dezful, which could raise water 50 cubits in height for the water supply to all houses in the town. Also diversion dams were known. Milling dams were introduced which the Muslim engineers called the *Pul-i-Bulaiti*. The first was built at Shustar on the River Karun, Iran, and many of these were later built in other parts of the Islamic world. Water was conducted from the back of the dam through a large pipe to drive a water wheel and watermill. In the 10th century, Al-Muqaddasi described several dams in Persia. He reported that one in Ahwaz was more than 910 m (3,000 ft) long, and that it had many water-wheels raising the water into aqueducts through which it flowed into reservoirs of the city. Another one, the Band-i-Amir dam, provided irrigation for 300 villages.

Middle Ages

In the Netherlands, a low-lying country, dams were often applied to block rivers in order to regulate the water level and to prevent the sea from entering the marsh lands. Such dams often marked the beginning of a town or city because it was easy to cross the river at such a place, and often gave rise to the respective place's names in Dutch.

For instance the Dutch capital Amsterdam (old name *Amstelredam*) started with a dam through the river Amstel in the late 12th century, and Rotterdam started with a dam through the river Rotte, a minor tributary of the Nieuwe Maas. The central square of Amsterdam, covering the original place of the 800-year-old dam, still carries the name *Dam Square* or simply *the Dam*.

Industrial Revolution

The Romans were the first to build arch dams, where the reaction forces from the abutment stabilizes the structure from the external hydrostatic pressure, but it was only in the 19th century that the engineering skills and construction materials available were capable of building the first large-scale arch dams.

Three pioneering arch dams were built around the British Empire in the early 19th century. Henry

Russel of the Royal Engineers oversaw the construction of the Mir Alam dam in 1804 to supply water to the city of Hyderabad (it is still in use today). It had a height of 12 m (39 ft) and consisted of 21 arches of variable span.

An engraving of the Rideau Canal locks at Bytown

In the 1820s and 30s, Lieutenant-Colonel John By supervised the construction of the Rideau Canal in Canada near modern-day Ottawa and built a series of curved masonry dams as part of the waterway system. In particular, the Jones Falls Dam, built by John Redpath, was completed in 1832 as the largest dam in North America and an engineering marvel. In order to keep the water in control during construction, two sluices, artificial channels for conducting water, were kept open in the dam. The first was near the base of the dam on its east side. A second sluice was put in on the west side of the dam, about 20 ft (6.1 m) above the base. To make the switch from the lower to upper sluice, the outlet of Sand Lake was blocked off.

Masonry arch wall, Parramatta, New South Wales, the first engineered dam built in Australia

Hunts Creek near the city of Parramatta, Australia, was dammed in the 1850s, to cater for the demand for water from the growing population of the city. The masonry arch dam wall was designed by Lieutenant Percy Simpson who was influenced by the advances in dam engineering techniques made by the Royal Engineers in India. The dam cost £17,000 and was completed in 1856 as the first engineered dam built in Australia, and the second arch dam in the world built to mathematical specifications.

The first such dam was opened two years earlier in France. It was the first French arch dam of the industrial era, and it was built by François Zola in the municipality of Aix-en-Provence to improve the supply of water after the 1832 cholera outbreak devastated the area. After royal approval was granted in 1844, the dam was constructed over the following decade. Its construction was carried out on the basis of the mathematical results of scientific stress analysis.

The 75-miles dam near Warwick, Australia, was possibly the world's first concrete arch dam. Designed by Henry Charles Stanley in 1880 with an overflow spillway and a special water outlet, it was eventually heightened to 10 m (33 ft).

In the latter half of the nineteenth century, significant advances in the scientific theory of masonry dam design were made. This transformed dam design from an art based on empirical methodology to a profession based on a rigorously applied scientific theoretical framework. This new emphasis was centered around the engineering faculties of universities in France and in the United Kingdom. William John Macquorn Rankine at the University of Glasgow pioneered the theoretical understanding of dam structures in his 1857 paper *On the Stability of Loose Earth*. Rankine theory provided a good understanding of the principles behind dam design. In France, J. Augustin Tortene de Sazilly explained the mechanics of vertically faced masonry gravity dams, and Zola's dam was the first to be built on the basis of these principles.

Large Dams

The Hoover Dam by Ansel Adams, 1942

The era of large dams was initiated with the construction of the Aswan Low Dam in Egypt in 1902, a gravity masonry buttress dam on the Nile River. Following their 1882 invasion and occupation of Egypt, the British began construction in 1898. The project was designed by Sir William Willcocks and involved several eminent engineers of the time, including Sir Benjamin Baker and Sir John Aird, whose firm, John Aird & Co., was the main contractor. Capital and financing were furnished by Ernest Cassel. When initially constructed between 1899 and 1902, nothing of its scale had ever been attempted; on completion, it was the largest masonry dam in the world.

The Hoover Dam is a massive concrete arch-gravity dam, constructed in the Black Canyon of the Colorado River, on the border between the US states of Arizona and Nevada between 1931 and 1936 during the Great Depression. In 1928, Congress authorized the project to build a dam that would control floods, provide irrigation water and produce hydroelectric power. The winning bid to build the dam was submitted by a consortium called Six Companies, Inc. Such a large concrete structure had never been built before, and some of the techniques were unproven. The torrid summer weather and the lack of facilities near the site also presented difficulties. Nevertheless, Six Companies turned over the dam to the federal government on 1 March 1936, more than two years ahead of schedule.

By 1997, there were an estimated 800,000 dams worldwide, some 40,000 of them over 15 m (49 ft) high. In 2014, scholars from the University of Oxford published a study of the cost of large dams – based on the largest existing dataset – documenting significant cost overruns for a majority of dams and questioning whether benefits typically offset costs for such dams.

Types of Dams

Dams can be formed by human agency, natural causes, or even by the intervention of wildlife such as beavers. Man-made dams are typically classified according to their size (height), intended purpose or structure.

By Structure

Based on structure and material used, dams are classified as easily created without materials, arch-gravity dams, embankment dams or masonry dams, with several subtypes.

Arch Dams

Gordon Dam, Tasmania, is an arch dam.

In the arch dam, stability is obtained by a combination of arch and gravity action. If the upstream face is vertical the entire weight of the dam must be carried to the foundation by gravity, while the distribution of the normal hydrostatic pressure between vertical cantilever and arch action will depend upon the stiffness of the dam in a vertical and horizontal direction. When the upstream face is sloped the distribution is more complicated. The normal component of the weight of the arch ring may be taken by the arch action, while the normal hydrostatic pressure will be distributed as described above. For this type of dam, firm reliable supports at the abutments (either buttress or canyon side wall) are more important. The most desirable place for an arch dam is a narrow canyon with steep side walls composed of sound rock. The safety of an arch dam is dependent on the strength of the side wall abutments, hence not only should the arch be well seated on the side walls but also the character of the rock should be carefully inspected.

Daniel-Johnson Dam, Quebec, is a multiple-arch buttress dam.

Two types of single-arch dams are in use, namely the constant-angle and the constant-radius dam. The constant-radius type employs the same face radius at all elevations of the dam, which means that as the channel grows narrower towards the bottom of the dam the central angle subtended by the face of the dam becomes smaller. Jones Falls Dam, in Canada, is a constant radius dam. In a constant-angle dam, also known as a variable radius dam, this subtended angle is kept a constant and the variation in distance between the abutments at various levels are taken care of by varying the radii. Constant-radius dams are much less common than constant-angle dams. Parker Dam on the Colorado River is a constant-angle arch dam.

A similar type is the double-curvature or thin-shell dam. Wildhorse Dam near Mountain City, Nevada, in the United States is an example of the type. This method of construction minimizes the amount of concrete necessary for construction but transmits large loads to the foundation and abutments. The appearance is similar to a single-arch dam but with a distinct vertical curvature to it as well lending it the vague appearance of a concave lens as viewed from downstream.

The multiple-arch dam consists of a number of single-arch dams with concrete buttresses as the supporting abutments, as for example the Daniel-Johnson Dam, Québec, Canada. The multiple-arch dam does not require as many buttresses as the hollow gravity type, but requires good rock foundation because the buttress loads are heavy.

Gravity Dams

In a gravity dam, the force that holds the dam in place against the push from the water is Earth's gravity pulling down on the mass of the dam. The water presses laterally (downstream) on the dam, tending to overturn the dam by rotating about its toe (a point at the bottom downstream side of the dam). The dam's weight counteracts that force, tending to rotate the dam the other way about its toe. The designer ensures that the dam is heavy enough that the dam's weight wins that contest. In engineering terms, that is true whenever the resultant of the forces of gravity acting on the dam and water pressure on the dam acts in a line that passes upstream of the toe of the dam.

The Grand Coulee Dam is an example of a solid gravity dam.

Furthermore, the designer tries to shape the dam so if one were to consider the part of dam above any particular height to be a whole dam itself, that dam also would be held in place by gravity. i.e. there is no tension in the upstream face of the dam holding the top of the dam down. The designer does this because it is usually more practical to make a dam of material essentially just piled up than to make the material stick together against vertical tension.

Note that the shape that prevents tension in the upstream face also eliminates a balancing compression stress in the downstream face, providing additional economy.

For this type of dam, it is essential to have an impervious foundation with high bearing strength.

When situated on a suitable site, a gravity dam can prove to be a better alternative to other types of dams. When built on a carefully studied foundation, the gravity dam probably represents the best developed example of dam building. Since the fear of flood is a strong motivator in many regions, gravity dams are being built in some instances where an arch dam would have been more economical.

Gravity dams are classified as "solid" or "hollow" and are generally made of either concrete or masonry. The solid form is the more widely used of the two, though the hollow dam is frequently more economical to construct. Grand Coulee Dam is a solid gravity dam and Braddock Locks & Dam is a hollow gravity dam.

Arch-gravity Dams

A gravity dam can be combined with an arch dam into an arch-gravity dam for areas with massive amounts of water flow but less material available for a purely gravity dam. The inward compression of the dam by the water reduces the lateral (horizontal) force acting on the dam. Thus, the gravitation force required by the dam is lessened, i.e. the dam does not need to be so massive. This enables thinner dams and saves resources.

The Hoover Dam is an example of an arch-gravity dam.

Barrages

The Koshi Barrage

A barrage dam is a special kind of dam which consists of a line of large gates that can be opened or closed to control the amount of water passing the dam. The gates are set between flanking piers which are responsible for supporting the water load, and are often used to control and stabilize water flow for irrigation systems. An example of this type of dam is the now-decommissioned Red Bluff Diversion Dam on the Sacramento River near Red Bluff, California.

Barrages that are built at the mouths of rivers or lagoons to prevent tidal incursions or utilize the tidal flow for tidal power are known as tidal barrages.

Embankment Dams

Embankment dams are made from compacted earth, and have two main types, rock-fill and earth-fill dams. Embankment dams rely on their weight to hold back the force of water, like gravity dams made from concrete.

Rock-fill Dams

Rock-fill dams are embankments of compacted free-draining granular earth with an impervious zone. The earth utilized often contains a high percentage of large particles, hence the term "rock-fill". The impervious zone may be on the upstream face and made of masonry, concrete, plastic membrane, steel sheet piles, timber or other material. The impervious zone may also be within the embankment in which case it is referred to as a *core*. In the instances where clay is utilized as the impervious material the dam is referred to as a *composite* dam. To prevent internal erosion of clay into the rock fill due to seepage forces, the core is separated using a filter. Filters are specifically graded soil designed to prevent the migration of fine grain soil particles. When suitable material is at hand, transportation is minimized leading to cost savings during construction. Rock-fill dams are resistant to damage from earthquakes. However, inadequate quality control during construction can lead to poor compaction and sand in the embankment which can lead to liquefaction of the rock-fill during an earthquake. Liquefaction potential can be reduced by keeping susceptible material from being saturated, and by providing adequate compaction during construction. An example of a rock-fill dam is New Melones Dam in California or the Fierza Dam in Albania.

The Gathright Dam in Virginia is a rock-fill embankment dam.

A core that is growing in popularity is asphalt concrete. The majority of such dams are built with rock and/or gravel as the main fill material. Almost 100 dams of this design have now been built worldwide since the first such dam was completed in 1962. All asphalt-concrete core dams built so far have an excellent performance record. The type of asphalt used is a viscoelastic-plastic material that can adjust to the movements and deformations imposed on the embankment as a whole, and to settlements in the foundation. The flexible properties of the asphalt make such dams especially suited in earthquake regions.

For the Moglicë Hydro Power Plant in Albania the Norwegian power company Statkraft is currently building an asphalt-core rock-fill dam. Upon completion in 2018 the 320 m long, 150 m high and 460 m wide dam is anticipated to be the world's highest of its kind.

Concrete-face Rock-fill Dams

A concrete-face rock-fill dam (CFRD) is a rock-fill dam with concrete slabs on its upstream face. This design provides the concrete slab as an impervious wall to prevent leakage and also a structure without concern for uplift pressure. In addition, the CFRD design is flexible for topography, faster to construct and less costly than earth-fill dams. The CFRD concept originated during the California Gold Rush in the 1860s when miners constructed rock-fill timber-face dams for sluice operations. The timber was later replaced by concrete as the design was applied to irrigation and power schemes. As CFRD designs grew in height during the 1960s, the fill was compacted and the slab's horizontal and vertical joints were replaced with improved vertical joints. In the last few decades, the design has become popular.

Currently, the tallest CFRD in the world is the 233 m-tall (764 ft) Shuibuya Dam in China which was completed in 2008.

Earth-fill Dams

Earth-fill dams, also called earthen dams, rolled-earth dams or simply earth dams, are constructed as a simple embankment of well compacted earth. A homogeneous rolled-earth dam is entirely constructed of one type of material but may contain a drain layer to collect seep water. A zoned-earth dam has distinct parts or zones of dissimilar material, typically a locally plentiful shell with a watertight clay core. Modern zoned-earth embankments employ filter and drain zones to collect and remove seep water and preserve the integrity of the downstream shell zone. An outdated method of zoned earth dam construction utilized a hydraulic fill to produce a watertight core. Rolled-earth dams may also employ a watertight facing or core in the manner of a rock-fill dam. An interesting type of temporary earth dam occasionally used in high latitudes is the frozen-core dam, in which a coolant is circulated through pipes inside the dam to maintain a watertight region of permafrost within it.

Tarbela Dam is a large dam on the Indus River in Pakistan. It is located about 50 km (31 mi) northwest of Islamabad, and a height of 485 ft (148 m) above the river bed and a reservoir size of 95 sq mi (250 km²) makes it the largest earth-filled dam in the world. The principal element of the project is an embankment 9,000 feet (2,700 m) long with a maximum height of 465 feet (142 m). The total volume of earth and rock used for the project is approximately 200 million cubic yards (152.8 million cu. meters) which makes it one of the largest man-made structures in the world.

Because earthen dams can be constructed from materials found on-site or nearby, they can be very cost-effective in regions where the cost of producing or bringing in concrete would be prohibitive.

By Size

International standards (including the International Commission on Large Dams, ICOLD) define *large dams* as higher than 15 m (49 ft) and *major dams* as over 150 m (490 ft) in height. The *Report of the World Commission on Dams* also includes in the *large* category, dams, such as barrages, which are between 5 and 15 m (16 and 49 ft) high with a reservoir capacity of more than 3 million cubic metres (2,400 acre·ft).

The tallest dam in the world is the 300 m-high (980 ft) Nurek Dam in Tajikistan.

By Use

Saddle Dam

A saddle dam is an auxiliary dam constructed to confine the reservoir created by a primary dam either to permit a higher water elevation and storage or to limit the extent of a reservoir for increased efficiency. An auxiliary dam is constructed in a low spot or "saddle" through which the reservoir would otherwise escape. On occasion, a reservoir is contained by a similar structure called a dike to prevent inundation of nearby land. Dikes are commonly used for reclamation of arable land from a shallow lake. This is similar to a levee, which is a wall or embankment built along a river or stream to protect adjacent land from flooding.

Weir

A weir (also sometimes called an *overflow dam*) is a type of small overflow dam that is often used within a river channel to create an impoundment lake for water abstraction purposes and which can also be used for flow measurement or retardation.

Check Dam

A check dam is a small dam designed to reduce flow velocity and control soil erosion. Conversely, a wing dam is a structure that only partly restricts a waterway, creating a faster channel that resists the accumulation of sediment.

Dry Dam

A dry dam, also known as a flood retarding structure, is a dam designed to control flooding. It normally holds back no water and allows the channel to flow freely, except during periods of intense flow that would otherwise cause flooding downstream.

Diversionary Dam

A diversionary dam is a structure designed to divert all or a portion of the flow of a river from its natural course. The water may be redirected into a canal or tunnel for irrigation and/or hydroelectric power production.

Underground Dam

Underground dams are used to trap groundwater and store all or most of it below the surface for extended use in a localized area. In some cases they are also built to prevent saltwater from intruding into a freshwater aquifer. Underground dams are typically constructed in areas where water resources are minimal and need to be efficiently stored, such as in deserts and on islands like the Fukuzato Dam in Okinawa, Japan. They are most common in northeastern Africa and the arid areas of Brazil while also being used in the southwestern United States, Mexico, India, Germany, Italy, Greece, France and Japan.

There are two types of underground dams: a *sub-surface* and a *sand-storage* dam. A sub-surface dam is built across an aquifer or drainage route from an impervious layer (such as solid bedrock) up to just below the surface. They can be constructed of a variety of materials to include bricks, stones, concrete, steel or PVC. Once built, the water stored behind the dam raises the water table and is then extracted with wells. A sand-storage dam is a weir built in stages across a stream or wadi. It must be strong, as floods will wash over its crest. Over time, sand accumulates in layers behind the dam, which helps store water and, most importantly, prevent evaporation. The stored water can be extracted with a well, through the dam body, or by means of a drain pipe.

Tailings Dam

A tailings dam is typically an earth-fill embankment dam used to store tailings, which are produced during mining operations after separating the valuable fraction from the uneconomic fraction of an ore. Conventional water retention dams can serve this purpose, but due to cost, a tailings dam is more viable. Unlike water retention dams, a tailings dam is raised in succession throughout the life of the particular mine. Typically, a base or starter dam is constructed, and as it fills with a mixture of tailings and water, it is raised. Material used to raise the dam can include the tailings (depending on their size) along with dirt.

There are three raised tailings dam designs, theee *upstream*, *downstream* and *centerline*, named according to the movement of the crest during raising. The specific design used is dependent upon topography, geology, climate, the type of tailings, and cost. An upstream tailings dam consists of trapezoidal embankments being constructed on top but toe to crest of another, moving the crest further upstream. This creates a relatively flat downstream side and a jagged upstream side which is supported by tailings slurry in the impoundment. The downstream design refers to the successive raising of the embankment that positions the fill and crest further downstream. A centerlined dam has sequential embankment dams constructed directly on top of another while fill is placed on the downstream side for support and slurry supports the upstream side.

Because tailings dams often store toxic chemicals from the mining process, they have an impervious liner to prevent seepage. Water/slurry levels in the tailings pond must be managed for stability and environmental purposes as well.

By Material

Steel Dams

Red Ridge steel dam, built 1905, Michigan

A steel dam is a type of dam briefly experimented with around the start of the 20th century which uses steel plating (at an angle) and load-bearing beams as the structure. Intended as permanent structures, steel dams were an (arguably failed) experiment to determine if a construction technique could be devised that was cheaper than masonry, concrete or earthworks, but sturdier than timber crib dams.

Timber Dams

A timber crib dam in Michigan, photographed in 1978

Timber dams were widely used in the early part of the industrial revolution and in frontier areas due to ease and speed of construction. Rarely built in modern times because of their relatively short lifespan and the limited height to which they can be built, timber dams must be kept constantly wet in order to maintain their water retention properties and limit deterioration by rot, similar to a barrel. The locations where timber dams are most economical to build are those where timber is plentiful, cement is costly or difficult to transport, and either a low head diversion dam is required or longevity is not an issue. Timber dams were once numerous, especially in the North American West, but most have failed, been hidden under earth embankments, or been replaced with entirely new structures. Two common variations of timber dams were the *crib* and the *plank*.

Timber crib dams were erected of heavy timbers or dressed logs in the manner of a log house and the interior filled with earth or rubble. The heavy crib structure supported the dam's face and the weight of the water. Splash dams were timber crib dams used to help float logs downstream in the late 19th and early 20th centuries.

Timber plank dams were more elegant structures that employed a variety of construction methods utilizing heavy timbers to support a water retaining arrangement of planks.

Other Types

Cofferdams

A cofferdam during the construction of locks at the Montgomery Point Lock and Dam

A cofferdam is a barrier, usually temporary, constructed to exclude water from an area that is normally submerged. Made commonly of wood, concrete, or steel sheet piling, cofferdams are used to allow construction on the foundation of permanent dams, bridges, and similar structures. When the project is completed, the cofferdam will usually be demolished or removed unless the area requires continuous maintenance.

Common uses for cofferdams include construction and repair of offshore oil platforms. In such cases the cofferdam is fabricated from sheet steel and welded into place under water. Air is pumped into the space, displacing the water and allowing a dry work environment below the surface.

Natural Dams

Dams can also be created by natural geological forces. Volcanic dams are formed when lava flows, often basaltic, intercept the path of a stream or lake outlet, resulting in the creation of a natural impoundment. An example would be the eruptions of the Uinkaret volcanic field about 1.8 million–10,000 years ago, which created lava dams on the Colorado River in northern Arizona in the United States. The largest such lake grew to about 800 km (500 mi) in length before the failure of its dam. Glacial activity can also form natural dams, such as the damming of the Clark Fork in Montana by the Cordilleran Ice Sheet, which formed the 7,780 km² (3,000 sq mi) Glacial Lake Missoula near the end of the last Ice Age. Moraine deposits left behind by glaciers can also dam rivers to form lakes, such as at Flathead Lake, also in Montana.

Natural disasters such as earthquakes and landslides frequently create landslide dams in mountainous regions with unstable local geology. Historical examples include the Usoi Dam in Tajikistan, which blocks the Murghab River to create Sarez Lake. At 560 m (1,840 ft) high, it is the tallest dam in the world, including both natural and man-made dams. A more recent example would be the creation of Attabad Lake by a landslide on Pakistan's Hunza River.

Natural dams often pose significant hazards to human settlements and infrastructure. The resulting lakes often flood inhabited areas, while a catastrophic failure of the dam could cause even greater damage, such as the failure of western Wyoming's Gros Ventre landslide dam in 1927, which wiped out the town of Kelly and resulted in the deaths of six people.

Beaver Dams

Beavers create dams primarily out of mud and sticks to flood a particular habitable area. By flooding a parcel of land, beavers can navigate below or near the surface and remain relatively well hidden or protected from predators. The flooded region also allows beavers access to food, especially during the winter.

Construction Elements

Power Generation Plant

As of 2005, hydroelectric power, mostly from dams, supplies some 19% of the world's electricity, and over 63% of renewable energy. Much of this is generated by large dams, although China uses small-scale hydro generation on a wide scale and is responsible for about 50% of world use of this type of power.

Most hydroelectric power comes from the potential energy of dammed water driving a water turbine and generator; to boost the power generation capabilities of a dam, the water may be run through a large pipe called a penstock before the turbine. A variant on this simple model uses pumped-storage hydroelectricity to produce electricity to match periods of high and low demand, by moving water between reservoirs at different elevations. At times of low electrical demand, excess generation capacity is used to pump water into the higher reservoir. When there is higher demand, water is released back into the lower reservoir through a turbine.

Spillways

Spillway on Llyn Brianne dam, Wales, soon after first fill

A spillway is a section of a dam designed to pass water from the upstream side of a dam to the downstream side. Many spillways have floodgates designed to control the flow through the spillway. There are several types of spillway. A *service spillway* or *primary spillway* passes normal flow. An *auxiliary spillway* releases flow in excess of the capacity of the service spillway. An *emergency spillway* is designed for extreme conditions, such as a serious malfunction of the service spillway. A *fuse plug spillway* is a low embankment designed to be overtopped and washed away in the event of a large flood. The elements of a fuse plug are independent free-standing blocks, set side by side which work without any remote control. They allow increasing the normal pool of the dam without compromising the security of the dam because they are designed to be gradually evacuated for exceptional events. They work as fixed weirs at times by allowing over-flow for common floods.

The spillway can be gradually eroded by water flow, including cavitation or turbulence of the water flowing over the spillway, leading to its failure. It was the inadequate design of the spillway which led to the 1889 over-topping of the South Fork Dam in Johnstown, Pennsylvania, resulting in the infamous Johnstown Flood (the "great flood of 1889").

Erosion rates are often monitored, and the risk is ordinarily minimized, by shaping the downstream face of the spillway into a curve that minimizes turbulent flow, such as an ogee curve.

Dam Creation

Common Purposes

Function	Example
Power generation	Hydroelectric power is a major source of electricity in the world. Many countries have rivers with adequate water flow, that can be dammed for power generation purposes. For example, the Itaipu Dam on the Paraná River in South America generates 14 GW and supplied 93% of the energy consumed by Paraguay and 20% of that consumed by Brazil as of 2005.
Water supply	Many urban areas of the world are supplied with water abstracted from rivers pent up behind low dams or weirs. Examples include London, with water from the River Thames, and Chester, with water taken from the River Dee. Other major sources include deep upland reservoirs contained by high dams across deep valleys, such as the Claerwen series of dams and reservoirs.
Stabilize water flow / irrigation	Dams are often used to control and stabilize water flow, often for agricultural purposes and irrigation. Others such as the Berg Strait dam can help to stabilize or restore the water levels of inland lakes and seas, in this case the Aral Sea.
Flood prevention	The Keenleyside Dam on the Columbia River, Canada can store 8.76 km³ (2.10 cu mi) of floodwaters, and the huge Delta Works protects the Netherlands from coastal flooding.
Land reclamation	Dams (often called dykes or levees in this context) are used to prevent ingress of water to an area that would otherwise be submerged, allowing its reclamation for human use.
Water diversion	A typically small dam used to divert water for irrigation, power generation, or other uses, with usually no other function. Occasionally, they are used to divert water to another drainage or reservoir to increase flow there and improve water use in that particular area. See: diversion dam.
Navigation	Dams create deep reservoirs and can also vary the flow of water downstream. This can in return affect upstream and downstream navigation by altering the river's depth. Deeper water increases or creates freedom of movement for water vessels. Large dams can serve this purpose, but most often weirs and locks are used.

Some of these purposes are conflicting, and the dam operator needs to make dynamic tradeoffs. For example, power generation and water supply would keep the reservoir high, whereas flood prevention would keep it low. Many dams in areas where precipitation fluctuates in an annual cycle will also see the reservoir fluctuate annually in an attempt to balance these difference purposes. Dam management becomes a complex exercise amongst competing stakeholders.

Location

The discharge of Takato Dam

One of the best places for building a dam is a narrow part of a deep river valley; the valley sides then can act as natural walls. The primary function of the dam's structure is to fill the gap in the natural reservoir line left by the stream channel. The sites are usually those where the gap becomes a minimum for the required storage capacity. The most economical arrangement is often a composite structure such as a masonry dam flanked by earth embankments. The current use of the land to be flooded should be dispensable.

Significant other engineering and engineering geology considerations when building a dam include:

- Permeability of the surrounding rock or soil
- Earthquake faults
- Landslides and slope stability
- Water table
- Peak flood flows
- Reservoir silting
- Environmental impacts on river fisheries, forests and wildlife
- Impacts on human habitations
- Compensation for land being flooded as well as population resettlement
- Removal of toxic materials and buildings from the proposed reservoir area

Impact Assessment

Impact is assessed in several ways: the benefits to human society arising from the dam (agriculture, water, damage prevention and power), harm or benefit to nature and wildlife, impact on the geology of an area (whether the change to water flow and levels will increase or decrease stability), and the disruption to human lives (relocation, loss of archeological or cultural matters underwater).

Environmental Impact

Wood and garbage accumulation due to a dam

Reservoirs held behind dams affect many ecological aspects of a river. Rivers topography and dynamics depend on a wide range of flows, whilst rivers below dams often experience long periods of very stable flow conditions or sawtooth flow patterns caused by releases followed by no releases. Water releases from a reservoir including that exiting a turbine usually contain very little suspended sediment, and this in turn can lead to scouring of river beds and loss of riverbanks; for example, the daily cyclic flow variation caused by the Glen Canyon Dam was a contributor to sand bar erosion.

Older dams often lack a fish ladder, which keeps many fish from moving upstream to their natural breeding grounds, causing failure of breeding cycles or blocking of migration paths. Even the presence of a fish ladder does not always prevent a reduction in fish reaching the spawning grounds upstream. In some areas, young fish ("smolt") are transported downstream by barge during parts of the year. Turbine and power-plant designs that have a lower impact upon aquatic life are an active area of research.

A large dam can cause the loss of entire ecospheres, including endangered and undiscovered species in the area, and the replacement of the original environment by a new inland lake.

Large reservoirs formed behind dams have been indicated in the contribution of seismic activity, due to changes in water load and/or the height of the water table.

Dams are also found to have a role in the increase/decrease of global warming. The changing water levels in reservoirs are a source for greenhouse gases like methane. While dams and the water behind them cover only a small portion of earth's surface, they harbour biological activity that can produce large amounts of greenhouse gases.

Human Social Impact

The impact on human society is also significant. Nick Cullather argues in *Hungry World: America's Cold War Battle Against Poverty in Asia* that dam construction requires the state to displace individual people in the name of the common good, and that it often leads to abuses of the masses by planners. He cites Morarji Desai, Interior Minister of India, in 1960 speaking to villagers upset about the Pong Dam, who threatened to "release the waters" and drown the villagers if they did not cooperate.

For example, the Three Gorges Dam on the Yangtze River in China is more than five times the size of the Hoover Dam (U.S.), and creates a reservoir 600 km (370 mi) long to be used for flood control and hydro-power generation. Its construction required the loss of over a million people's homes and their mass relocation, the loss of many valuable archaeological and cultural sites, as well as significant ecological change. During the 2010 China floods, the dam held back a what would have been a disastrous flood and the huge reservoir rose by 4 m (13 ft) overnight.

It is estimated that to date, 40–80 million people worldwide have been physically displaced from their homes as a result of dam construction.

Economics

Construction of a hydroelectric plant requires a long lead time for site studies, hydrological studies, and environmental impact assessments, and are large-scale projects by comparison to traditional power generation based upon fossil fuels. The number of sites that can be economically developed for hydroelectric production is limited; new sites tend to be far from population centers and usually require extensive power transmission lines. Hydroelectric generation can be vulnerable to major changes in the climate, including variations in rainfall, ground and surface water levels, and glacial melt, causing additional expenditure for the extra capacity to ensure sufficient power is available in low-water years.

Once completed, if it is well designed and maintained, a hydroelectric power source is usually comparatively cheap and reliable. It has no fuel and low escape risk, and as an alternative energy source it is cheaper than both nuclear and wind power. It is more easily regulated to store water as needed and generate high power levels on demand compared to wind power.

Dam Failure

The reservoir emptying through the failed Teton Dam

International special sign for works and installations containing dangerous forces

Dam failures are generally catastrophic if the structure is breached or significantly damaged. Routine deformation monitoring and monitoring of seepage from drains in and around larger dams is useful to anticipate any problems and permit remedial action to be taken before structural failure occurs. Most dams incorporate mechanisms to permit the reservoir to be lowered or even drained in the event of such problems. Another solution can be rock grouting – pressure pumping portland cement slurry into weak fractured rock.

During an armed conflict, a dam is to be considered as an "installation containing dangerous forces" due to the massive impact of a possible destruction on the civilian population and the environment. As such, it is protected by the rules of international humanitarian law (IHL) and shall not be made the object of attack if that may cause severe losses among the civilian population. To facilitate the identification, a protective sign consisting of three bright orange circles placed on the same axis is defined by the rules of IHL.

The main causes of dam failure include inadequate spillway capacity, piping through the embankment, foundation or abutments, spillway design error (South Fork Dam), geological instability caused by changes to water levels during filling or poor surveying (Vajont, Malpasset, Testalinden Creek dams), poor maintenance, especially of outlet pipes (Lawn Lake Dam, Val di Stava Dam collapse), extreme rainfall (Shakidor Dam), earthquakes, and human, computer or design error (Buffalo Creek Flood, Dale Dike Reservoir, Taum Sauk pumped storage plant).

A notable case of deliberate dam failure (prior to the above ruling) was the Royal Air Force 'Dambusters' raid on Germany in World War II (codenamed "Operation Chastise"), in which three German dams were selected to be breached in order to damage German infrastructure and manufacturing and power capabilities deriving from the Ruhr and Eder rivers. This raid later became the basis for several films.

Since 2007, the Dutch IJkdijk foundation is developing, with an open innovation model and early warning system for levee/dike failures. As a part of the development effort, full-scale dikes are destroyed in the IJkdijk fieldlab. The destruction process is monitored by sensor networks from an international group of companies and scientific institutions.

Sediment Transport

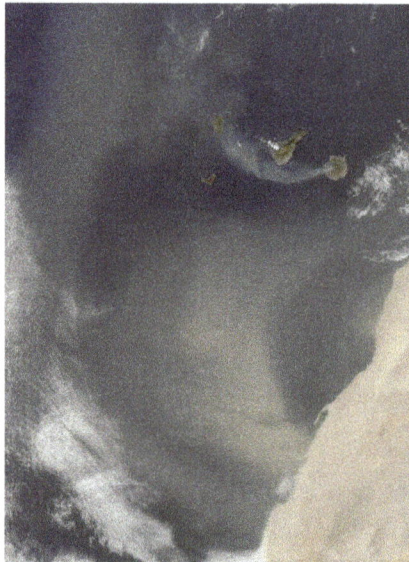

Dust blows from the Sahara Desert over the Atlantic Ocean towards the Canary Islands.

Sediment transport is the movement of solid particles (sediment), typically due to a combination of gravity acting on the sediment, and/or the movement of the fluid in which the sediment is entrained. Sediment transport occurs in natural systems where the particles are clastic rocks (sand, gravel, boulders, etc.), mud, or clay; the fluid is air, water, or ice; and the force of gravity acts to move the particles along the sloping surface on which they are resting. Sediment transport due to fluid motion occurs in rivers, oceans, lakes, seas, and other bodies of water due to currents and tides. Transport is also caused by glaciers as they flow, and on terrestrial surfaces under the influence of wind. Sediment transport due only to gravity can occur on sloping surfaces in general, including hillslopes, scarps, cliffs, and the continental shelf—continental slope boundary.

Sediment transport is important in the fields of sedimentary geology, geomorphology, civil engineering and environmental engineering. Knowledge of sediment transport is most often used to determine whether erosion or deposition will occur, the magnitude of this erosion or deposition, and the time and distance over which it will occur.

Mechanisms

Sand blowing off a crest in the Kelso Dunes of the Mojave Desert, California.

Toklat River, East Fork, Polychrome overlook, Denali National Park, Alaska. This river, like other braided streams, rapidly changes the positions of its channels through processes of erosion, sediment transport, and deposition.

Aeolian

Aeolian or *eolian* (depending on the parsing of æ) is the term for sediment transport by wind. This process results in the formation of ripples and sand dunes. Typically, the size of the transported sediment is fine sand (<1 mm) and smaller, because air is a fluid with low density and viscosity, and can therefore not exert very much shear on its bed.

Bedforms are generated by aeolian sediment transport in the terrestrial near-surface environment. Ripples and dunes form as a natural self-organizing response to sediment transport.

Aeolian sediment transport is common on beaches and in the arid regions of the world, because it is in these environments that vegetation does not prevent the presence and motion of fields of sand.

Wind-blown very fine-grained dust is capable of entering the upper atmosphere and moving across the globe. Dust from the Sahara deposits on the Canary Islands and islands in the Caribbean, and dust from the Gobi desert has deposited on the western United States. This sediment is important to the soil budget and ecology of several islands.

Deposits of fine-grained wind-blown glacial sediment are called loess.

Fluvial

In geology, physical geography, and sediment transport, fluvial processes relate to flowing water in natural systems. This encompasses rivers, streams, periglacial flows, flash floods and glacial lake outburst floods. Sediment moved by water can be larger than sediment moved by air because water has both a higher density and viscosity. In typical rivers the largest carried sediment is of sand and gravel size, but larger floods can carry cobbles and even boulders.

Fluvial sediment transport can result in the formation of ripples and dunes, in fractal-shaped patterns of erosion, in complex patterns of natural river systems, and in the development of floodplains.

Sand ripples, Laysan Beach, Hawaii. Coastal sediment transport results in these evenly spaced ripples along the shore. Monk seal for scale.

Coastal

Coastal sediment transport takes place in near-shore environments due to the motions of waves and currents. At the mouths of rivers, coastal sediment and fluvial sediment transport processes mesh to create river deltas.

Coastal sediment transport results in the formation of characteristic coastal landforms such as beaches, barrier islands, and capes.

A glacier joining the Gorner Glacier, Zermatt, Switzerland. These glaciers transport sediment and leave behind lateral moraines.

Glacial

As glaciers move over their beds, they entrain and move material of all sizes. Glaciers can carry the largest sediment, and areas of glacial deposition often contain a large number of glacial erratics, many of which are several metres in diameter. Glaciers also pulverize rock into "glacial flour", which is so fine that it is often carried away by winds to create loess deposits thousands of kilometres afield. Sediment entrained in glaciers often moves approximately along the glacial flowlines, causing it to appear at the surface in the ablation zone.

Hillslope

In hillslope sediment transport, a variety of processes move regolith downslope. These include:

- Soil creep

- Tree throw

- Movement of soil by burrowing animals

- Slumping and landsliding of the hillslope

These processes generally combine to give the hillslope a profile that looks like a solution to the diffusion equation, where the diffusivity is a parameter that relates to the ease of sediment transport on the particular hillslope. For this reason, the tops of hills generally have a parabolic concave-up profile, which grades into a convex-up profile around valleys.

As hillslopes steepen, however, they become more prone to episodic landslides and other mass wasting events. Therefore, hillslope processes are better described by a nonlinear diffusion equation in which classic diffusion dominates for shallow slopes and erosion rates go to infinity as the hillslope reaches a critical angle of repose.

Debris Flow

Large masses of material are moved in debris flows, hyperconcentrated mixtures of mud, clasts that range up to boulder-size, and water. Debris flows move as granular flows down steep mountain valleys and washes. Because they transport sediment as a granular mixture, their transport mechanisms and capacities scale differently from those of fluvial systems.

Applications

Sediment transport is applied to solve many environmental, geotechnical, and geological problems. Measuring or quantifying sediment transport or erosion is therefore important for coastal engineering. Several sediment erosion devices have been designed in order to quantitfy sediment erosion (e.g., Particle Erosion Simulator (PES)). One such device, also referred to as the BEAST (Benthic Environmental Assessment Sediment Tool) has been calibrated in order to quantify rates of sediment erosion.

Movement of sediment is important in providing habitat for fish and other organisms in rivers. Therefore, managers of highly regulated rivers, which are often sediment-starved due to dams, are often advised to stage short floods to refresh the bed material and rebuild bars. This is also important, for example, in the Grand Canyon of the Colorado River, to rebuild shoreline habitats also used as campsites.

Suspended sediment from a stream emptying into a fjord (Isfjorden, Svalbard, Norway).

Sediment discharge into a reservoir formed by a dam forms a reservoir delta. This delta will fill the basin, and eventually, either the reservoir will need to be dredged or the dam will need to be removed. Knowledge of sediment transport can be used to properly plan to extend the life of a dam.

Geologists can use inverse solutions of transport relationships to understand flow depth, velocity, and direction, from sedimentary rocks and young deposits of alluvial materials.

Flow in culverts, over dams, and around bridge piers can cause erosion of the bed. This erosion can damage the environment and expose or unsettle the foundations of the structure. Therefore, good knowledge of the mechanics of sediment transport in a built environment are important for civil and hydraulic engineers.

When suspended sediment transport is increased due to human activities, causing environmental problems including the filling of channels, it is called siltation after the grain-size fraction dominating the process.

Initiation of Motion

Stress Balance

For a fluid to begin transporting sediment that is currently at rest on a surface, the boundary (or bed) shear stress τ_b exerted by the fluid must exceed the critical shear stress τ_c for the initiation of motion of grains at the bed. This basic criterion for the initiation of motion can be written as:

$$\tau_b = \tau_c.$$

This is typically represented by a comparison between a dimensionless shear stress ($\tau_b{}^*$) and a dimensionless critical shear stress ($\tau_c{}^*$). The nondimensionalization is in order to compare the driving forces of particle motion (shear stress) to the resisting forces that would make it stationary (particle density and size). This dimensionless shear stress, τ^*, is called the Shields parameter and is defined as:

$$\tau^* = \frac{\tau}{(\rho_s - \rho_f)(g)(D)}.$$

And the new equation to solve becomes:

$$\tau_b{}^* \quad \tau_c{}^*.$$

The equations included here describe sediment transport for clastic, or granular sediment. They do not work for clays and muds because these types of floccular sediments do not fit the geometric simplifications in these equations, and also interact thorough electrostatic forces. The equations were also designed for fluvial sediment transport of particles carried along in a liquid flow, such as that in a river, canal, or other open channel.

Only one size of particle is considered in this equation. However, river beds are often formed by a mixture of sediment of various sizes. In case of partial motion where only a part of the sediment mixture moves, the river bed becomes enriched in large gravel as the smaller sediments are washed away. The smaller sediments present under this layer of large gravel have a lower possibility of movement and total sediment transport decreases. This is called armouring effect. Other forms of armouring of sediment or decreasing rates of sediment erosion can be caused by carpets of microbial mats, under conditions of high organic loading.

Critical Shear Stress

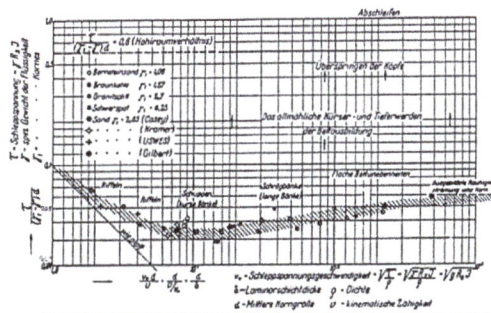

Original Shields diagram, 1936

The Shields diagram empirically shows how the dimensionless critical shear stress (i.e. the dimensionless shear stress required for the initiation of motion) is a function of a particular form of the particle Reynolds number, Re_p or Reynolds number related to the particle. This allows us to rewrite the criterion for the initiation of motion in terms of only needing to solve for a specific version of the particle Reynolds number, which we call $Re_p{}^*$.

$$\tau_b{}^* = f\left(Re_p{}^*\right)$$

This equation can then be solved by using the empirically derived Shields curve to $\tau_c{}^*$ find as a function of a specific form of the particle Reynolds number called the boundary Reynolds number. The mathematical solution of the equation was given by Dey.

Particle Reynolds Number

In general, a particle Reynolds Number has the form:

$$Re_p = \frac{U_p D}{v}$$

Where U_p is a characteristic particle velocity, D is the grain diameter (a characteristic particle size), and ν is the kinematic viscosity, which is given by the dynamic viscosity, μ, divided by the fluid density, ρ_f.

$$\nu = \frac{\mu}{\rho_f}$$

The specific particle Reynolds number of interest is called the boundary Reynolds number, and it is formed by replacing the velocity term in the Particle Reynolds number by the shear velocity, u_*, which is a way of rewriting shear stress in terms of velocity.

$$u_* = \sqrt{\frac{\tau_b}{\rho_f}} = \kappa z \frac{\partial u}{\partial z}$$

where τ_b is the bed shear stress (described below), and κ is the von Kármán constant, where

$$\kappa = 0.407.$$

The particle Reynolds number is therefore given by:

$$\mathrm{Re}_p{}^* = \frac{u_* D}{\nu}$$

Bed shear stress

The boundary Reynolds number can be used with the Shields diagram to empirically solve the equation

$$\tau_c{}^* = f\left(\mathrm{Re}_p{}^*\right),$$

which solves the right-hand side of the equation

$$\tau_b{}^* = \tau_c{}^*.$$

In order to solve the left-hand side, expanded as

$$\tau_b{}^* = \frac{\tau_b}{(\rho_s - \rho_f)(g)(D)},$$

we must find the bed shear stress, τ_b. There are several ways to solve for the bed shear stress. First, we develop the simplest approach, in which the flow is assumed to be steady and uniform and reach-averaged depth and slope are used. Due to the difficulty of measuring shear stress *in situ*, this method is also one of the most-commonly used. This method is known as the depth-slope product.

Depth-slope Product

For a river undergoing approximately steady, uniform equilibrium flow, of approximately constant depth h and slope angle θ over the reach of interest, and whose width is much greater than its depth, the bed shear stress is given by some momentum considerations stating that the gravity force component in the flow direction equals exactly the friction force. For a wide channel, it yields:

$$\tau_b = \rho g h \sin(\theta)$$

For shallow slope angles, which are found in almost all natural lowland streams, the small-angle formula shows that $\sin(\theta)$ is approximately equal to $\tan(\theta)$, which is given by S, the slope. Rewritten with this:

$$\tau_b = \rho g h S$$

Shear Velocity, Velocity, and Friction Factor

For the steady case, by extrapolating the depth-slope product and the equation for shear velocity:

$$\tau_b = \rho g h S$$

$$u_* = \sqrt{\left(\frac{\tau_b}{\rho}\right)},$$

We can see that the depth-slope product can be rewritten as:

$$\tau_b = \rho u_*^2$$

$u*$ is related to the mean flow velocity, $\bar{u}$, through the generalized Darcy-Weisbach friction factor, C_f, which is equal to the Darcy-Weisbach friction factor divided by 8 (for mathematical convenience). Inserting this friction factor,

$$\tau_b = \rho C_f \left(\bar{u}\right)^2.$$

Unsteady Flow

For all flows that cannot be simplified as a single-slope infinite channel (as in the depth-slope product, above), the bed shear stress can be locally found by applying the Saint-Vennant equations for continuity, which consider accelerations within the flow.

Example

Set-up

The criterion for the initiation of motion, established earlier, states that

$$\tau_b{}^* = \tau_c{}^*.$$

In this equation,

$$\tau^* = \frac{\tau}{(\rho_s - \rho)(g)(D)}, \text{ and therefore}$$

$$\frac{\tau_b}{(\rho_s - \rho)(g)(D)} = \frac{\tau_c}{(\rho_s - \rho)(g)(D)}.$$

τ_c^* is a function of boundary Reynolds number, a specific type of particle Reynolds number.

$$\tau_c^* = f\left(Re_p^*\right).$$

For a particular particle Reynolds number, τ_c^* will be an emprical constant given by the Shields Curve or by another set of empirical data (depending on whether or not the grain size is uniform).

Therefore, the final equation that we seek to solve is:

$$\frac{\tau_b}{(\rho_s - \rho)(g)(D)} = f\left(Re_p^*\right).$$

Solution

We make several assumptions to provide an example that will allow us to bring the above form of the equation into a solved form.

First, we assume that the a good approximation of reach-averaged shear stress is given by the depth-slope product. We can then rewrite the equation as

$$\rho g h S = 0.06(\rho_s - \rho)(g)(D).$$

Moving and re-combining the terms, we obtain:

$$hS = \frac{(\rho_s - \rho)}{\rho}(D)\left(f\left(\mathrm{Re}_p^*\right)\right) = RD\left(f\left(\mathrm{Re}_p^*\right)\right)$$

where R is the submerged specific gravity of the sediment.

We then make our second assumption, which is that the particle Reynolds number is high. This is typically applicable to particles of gravel-size or larger in a stream, and means that the critical shear stress is a constant. The Shields curve shows that for a bed with a uniform grain size,

$$\tau_c^* = 0.06.$$

Later researchers have shown that this value is closer to

$$\tau_c^* = 0.03$$

for more uniformly sorted beds. Therefore, we will simply insert

$$\tau_c^* = f\left(\mathrm{Re}_p^*\right)$$

and insert both values at the end.

The equation now reads:

$$hS = RD\tau_c^*$$

This final expression shows that the product of the channel depth and slope is equal to the Shield's criterion times the submerged specific gravity of the particles times the particle diameter.

For a typical situation, such as quartz-rich sediment $\left(\rho_s = 2650 \frac{kg}{m^3} \right)$ in water $\left(\rho = 1000 \frac{kg}{m^3} \right)$, the submerged specific gravity is equal to 1.65.

$$R = \frac{(\rho_s - \rho)}{\rho} = 1.65$$

Plugging this into the equation above,

$$hS = 1.65(D)\tau_c*.$$

For the Shield's criterion of $\tau_c* = 0.06$. 0.06 * 1.65 = 0.099, which is well within standard margins of error of 0.1. Therefore, for a uniform bed,

$$hS = 0.1(D).$$

For these situations, the product of the depth and slope of the flow should be 10% of the diameter of the median grain diameter.

The mixed-grain-size bed value is $\tau_c* = 0.03$, , which is supported by more recent research as being more broadly applicable because most natural streams have mixed grain sizes. Using this value, and changing D to D_50 ("50" for the 50th percentile, or the median grain size, as we are now looking at a mixed-grain-size bed), the equation becomes:

$$hS = 0.05(D_{50})$$

Which means that the depth times the slope should be about 5% of the median grain diameter in the case of a mixed-grain-size bed.

Modes of Entrainment

The sediments entrained in a flow can be transported along the bed as bed load in the form of sliding and rolling grains, or in suspension as suspended load advected by the main flow. Some sediment materials may also come from the upstream reaches and be carried downstream in the form of wash load.

Rouse Number

The location in the flow in which a particle is entrained is determined by the Rouse number, which is determined by the density ρ_s and diameter d of the sediment particle, and the density ρ and kinematic viscosity v of the fluid, determine in which part of the flow the sediment particle will be carried.

$$P = \frac{w_s}{\kappa u_*}$$

Here, the Rouse number is given by P. The term in the numerator is the (downwards) sediment the sediment settling velocity w_s, which is discussed below. The upwards velocity on the grain is given as a product of the von Kármán constant, $\kappa = 0.4$, and the shear velocity, u_*.

The following table gives the approximate required Rouse numbers for transport as bed load, suspended load, and wash load.

Mode of Transport	Rouse Number
Initiation of motion	>7.5
Bed load	>2.5, <7.5
Suspended load: 50% Suspended	>1.2, <2.5
Suspended load: 100% Suspended	>0.8, <1.2
Wash load	<0.8

Settling Velocity

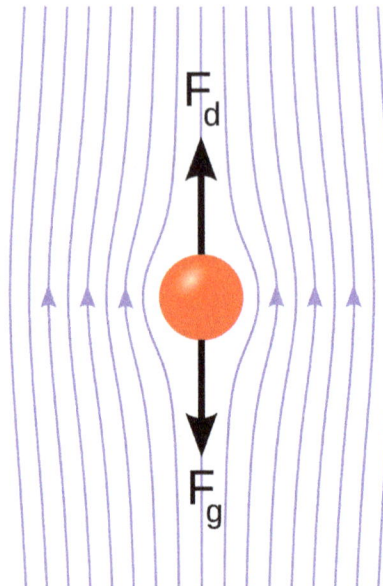

Streamlines around a sphere falling through a fluid. This illustration is accurate for laminar flow, in which the particle Reynolds number is small. This is typical for small particles falling through a viscous fluid; larger particles would result in the creation of a turbulent wake.

The settling velocity (also called the "fall velocity" or "terminal velocity") is a function of the particle Reynolds number. Generally, for small particles (laminar approximation), it can be calculated with Stokes' Law. For larger particles (turbulent particle Reynolds numbers), fall velocity is calculated with the turbulent drag law. Dietrich (1982) compiled a large amount of published data to which he empirically fit settling velocity curves. Ferguson and Church (2006) analytically combined the expressions for Stokes flow and a turbulent drag law into a single equation that works for all sizes of sediment, and successfully tested it against the data of Dietrich. Their equation is

$$w_s = \frac{RgD^2}{C_1 v + (0.75 C_2 RgD^3)^{(0.5)}}.$$

In this equation w_s is the sediment settling velocity, g is acceleration due to gravity, and D is mean sediment diameter. v is the kinematic viscosity of water, which is approximately 1.0 x 10^{-6} m²/s for water at 20 °C.

C_1 and C_2 are constants related to the shape and smoothness of the grains.

Constant	Smooth Spheres	Natural Grains: Sieve Diameters	Natural Grains: Nominal Diameters	Limit for Ultra-Angular Grains
C_1	18	18	20	24
C_2	0.4	1.0	1.1	1.2

The expression for fall velocity can be simplified so that it can be solved only in terms of D. We use the sieve diameters for natural grains, $g = 9.8,$, and values given above for v and R. From these parameters, the fall velocity is given by the expression:

$$w_s = \frac{16.17D^2}{1.8 \cdot 10^{-5} + (12.1275D^3)^{(0.5)}}$$

Hjulström-Sundborg Diagram

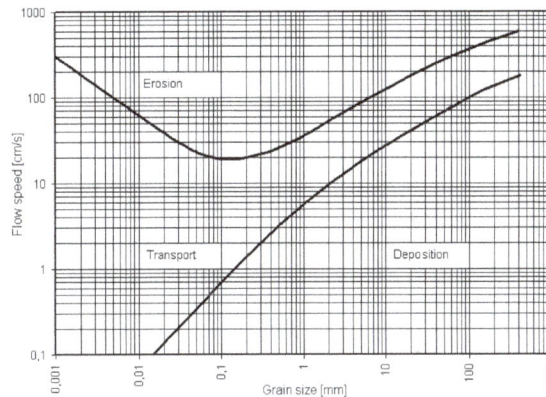

The logarithmic Hjulström curve

In 1935, Filip Hjulström created the Hjulström curve, a graph which shows the relationship between the size of sediment and the velocity required to erode (lift it), transport it, or deposit it. The graph is logarithmic.

Åke Sundborg later modified the Hjulström curve to show separate curves for the movement threshold corresponding to several water depths, as is necessary if the flow velocity rather than the boundary shear stress (as in the Shields diagram) is used for the flow strength.

This curve has no more than a historical value nowadays, although its simplicity is still attractive. Among the drawbacks of this curve are that it does not take the water depth into account and more importantly, that it does not show that sedimentation is caused by flow velocity *deceleration* and erosion is caused by flow *acceleration*. The dimensionless Shields diagram is now unanimously accepted for initiation of sediment motion in rivers. Much work was done on river sediment transport formulae in the second half of the 20th century and that work should be used preferably to Hjulström's curve, e.g. Meyer-Peter & Müller (1948), Engelund-Hansen (1967), Lefort (1991), Belleudy (2012).

Transport Rate

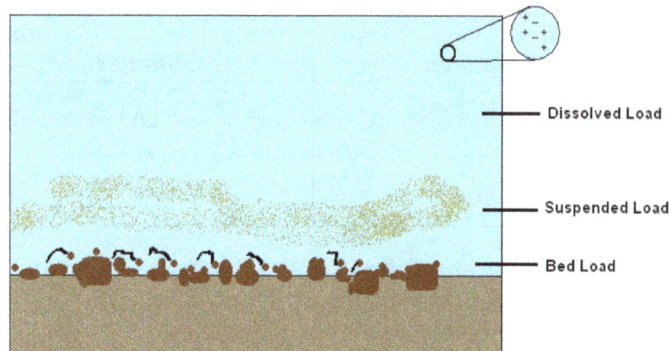

A schematic diagram of where the different types of sediment load are carried in the flow. Dissolved load is not sediment: it is composed of disassociated ions moving along with the flow. It may, however, constitute a significant proportion (often several percent, but occasionally greater than half) of the total amount of material being transported by the stream.

Formulas to calculate sediment transport rate exist for sediment moving in several different parts of the flow. These formulas are often segregated into bed load, suspended load, and wash load. They may sometimes also be segregated into bed material load and wash load.

Bed Load

Bed load moves by rolling, sliding, and hopping (or saltating) over the bed, and moves at a small fraction of the fluid flow velocity. Bed load is generally thought to constitute 5-10% of the total sediment load in a stream, making it less important in terms of mass balance. However, the bed material load (the bed load plus the portion of the suspended load which comprises material derived from the bed) is often dominated by bed load, especially in gravel-bed rivers. This bed material load is the only part of the sediment load that actively interacts with the bed. As the bed load is an important component of that, it plays a major role in controlling the morphology of the channel.

Bed load transport rates are usually expressed as being related to excess dimensionless shear stress raised to some power. Excess dimensionless shear stress is a nondimensional measure of bed shear stress about the threshold for motion.

$$\left(\tau_b^* - \tau_c^*\right),$$

Bed load transport rates may also be given by a ratio of bed shear stress to critical shear stress, which is equivalent in both the dimensional and nondimensional cases. This ratio is called the "transport stage" (T_s or ϕ) and is an important in that it shows bed shear stress as a multiple of the value of the criterion for the initiation of motion.

$$T_s = \phi = \frac{\tau_b}{\tau_c}$$

When used for sediment transport formulae, this ratio is typically raised to a power.

The majority of the published relations for bedload transport are given in dry sediment weight per unit channel width, b ("breadth"):

$$q_s = \frac{Q_s}{b}.$$

Due to the difficulty of estimating bed load transport rates, these equations are typically only suitable for the situations for which they were designed.

Notable Bed Load Transport Formulae

Meyer-Peter Müller and Derivatives

The transport formula of Meyer-Peter and Müller, originally developed in 1948, was designed for well-sorted fine gravel at a transport stage of about 8. The formula uses the above nondimensionalization for shear stress,

$$\tau^* = \frac{\tau}{(\rho_s - \rho)(g)(D)},$$

and Hans Einstein's nondimensionalization for sediment volumetric discharge per unit width

$$q_s^* = \frac{q_s}{D\sqrt{\frac{\rho_s - \rho}{\rho}gD}} = \frac{q_s}{Re_p v}.$$

Their formula reads:

$$q_s^* = 8\left(\tau^* - \tau^*_c\right)^{3/2}.$$

Their experimentally determined value for τ^*_c is 0.047, and is the third commonly used value for this (in addition to Parker's 0.03 and Shields' 0.06).

Because of its broad use, some revisions to the formula have taken place over the years that show that the coefficient on the left ("8" above) is a function of the transport stage:

$$T_s \approx 2 \rightarrow q_s^* = 5.7\left(\tau^* - 0.047\right)^{3/2}$$

$$T_s \approx 100 \rightarrow q_s^* = 12.1\left(\tau^* - 0.047\right)^{3/2[}$$

The variations in the coefficient were later generalized as a function of dimensionless shear stress:

$$\begin{cases} q_s^* = \alpha_s \left(\tau^* - \tau_c^*\right)^n \\ n = \frac{3}{2} \\ \alpha_s = 1.6\ln\left(\tau^*\right) + 9.8 \approx 9.64\tau^{*0.166} \end{cases}$$

Wilcock and Crowe

In 2003, Peter Wilcock and Joanna Crowe (now Joanna Curran) published a sediment transport formula that works with multiple grain sizes across the sand and gravel range. Their formula works with surface grain size distributions, as opposed to older models which use subsurface grain size distributions (and thereby implicitly infer a surface grain sorting).

Their expression is more complicated than the basic sediment transport rules (such as that of Meyer-Peter and Müller) because it takes into account multiple grain sizes: this requires consideration of reference shear stresses for each grain size, the fraction of the total sediment supply that falls into each grain size class, and a "hiding function".

The "hiding function" takes into account the fact that, while small grains are inherently more mobile than large grains, on a mixed-grain-size bed, they may be trapped in deep pockets between large grains. Likewise, a large grain on a bed of small particles will be stuck in a much smaller pocket than if it were on a bed of grains of the same size. In gravel-bed rivers, this can cause "equal mobility", in which small grains can move just as easily as large ones. As sand is added to the system, it moves away from the "equal mobility" portion of the hiding function to one in which grain size again matters.

Their model is based on the transport stage, or ratio of bed shear stress to critical shear stress for the initiation of grain motion. Because their formula works with several grain sizes simultaneously, they define the critical shear stress for each grain size class, τ_{c,D_i}, to be equal to a "reference shear stress", τ_{ri}..

They express their equations in terms of a dimensionless transport parameter, W_i^* (with the " * " indicating nondimensionality and the " $_i$ * " indicating that it is a function of grain size):

$$W_i^* = \frac{Rgq_{bi}}{F_i u *^3}$$

q_{bi} is the volumetric bed load transport rate of size class i per unit channel width b. F_i is the proportion of size class i that is present on the bed.

They came up with two equations, depending on the transport stage, ϕ.. For   1.35:

$$W_i^* = 0.002\phi^{7.5}$$

and for $\phi \geq 1.35$:

$$W_i^* = 14\left(1 - \frac{0.894}{\phi^{0.5}}\right)^{4.5}.$$

This equation asymptotically reaches a constant value of W_i^* as ϕ becomes large.

Wilcock and Kenworthy

In 2002, Peter Wilcock and Kenworthy T.A. , following Peter Wilcock (1998), published a sediment bed-load transport formula that works with only two sediments fractions, i.e. sand and gravel fractions. Peter Wilcock and Kenworthy T.A. in their article recognized that a mixed-sized sediment bed-load transport model using only two fractions offers practical advantages in terms of both computational and conceptual modeling by taking into account the nonlinear effects of sand presence in gravel beds on bed-load transport rate of both fractions. In fact, in the two-fraction bed load formula appears a new ingredient with respect to that of Meyer-Peter and Müller that is the proportion F_i of fraction i on the bed surface where the subscript $_i$ represents either the sand (s) or gravel (g) fraction. The proportion F_i, as a function of sand content f_s, physically represents the relative influence of the mechanisms controlling sand and gravel transport, associated with the change from a clast-supported to matrix-supported gravel bed. Moreover, since f_s spans between 0 and 1, phenomena that vary with f_s include the relative size effects producing "hiding" of fine grains and "exposure" of coarse grains. The "hiding" effect takes into account the fact that, while small grains are inherently more mobile than large grains, on a mixed-grain-size bed, they may be trapped in deep pockets between large grains. Likewise, a large grain on a bed of small particles will be stuck in a much smaller pocket than if it were on a bed of grains of the same size, which the Meyer-Peter and Müller formula refers to. In gravel-bed rivers, this can cause "equal mobility", in which small grains can move just as easily as large ones. As sand is added to the system, it moves away from the "equal mobility" portion of the hiding function to one in which grain size again matters.

Their model is based on the transport stage,i.e. ϕ, or ratio of bed shear stress to critical shear stress for the initiation of grain motion. Because their formula works with only two fractions simultaneously, they define the critical shear stress for each of the two grain size classes, τ_{ri}, where $_i$ represents either the sand (s) or gravel (g) fraction . The critical shear stress that represents the incipient motion for each of the two fractions is consistent with established values in the limit of pure sand and gravel beds and shows a sharp change with increasing sand content over the transition from a clast- to matrix-supported bed.

They express their equations in terms of a dimensionless transport parameter, W_i^* (with the "$*$" indicating nondimensionality and the "$_i$" indicating that it is a function of grain size):

$$W_i^* = \frac{Rgq_{bi}}{F_i u *^3}$$

q_{bi} is the volumetric bed load transport rate of size class i per unit channel width b. F_i is the proportion of size class i that is present on the bed.

They came up with two equations, depending on the transport stage, ϕ. For $\phi < \phi'$:

$$W_i^* = 0.002\phi^{7.5}$$

and for $\phi \geq \phi'$:

$$W_i^* = A\left(1 - \frac{\chi}{\phi^{0.5}}\right)^{4.5}.$$

This equation asymptotically reaches a constant value of W_i^*. as ϕ becomes large and the symbols A, ϕ, χ have the following values:

$$A = 70, \phi' = 1.19, \chi = 0.908, laboratory$$

$$A = 115, \phi' = 1.27, \chi = 0.923, field$$

In order to apply the above formulation, it is necessary to specify the characteristic grain sizes D_s for the sand portion and D_g for the gravel portion of the surface layer, the fractions F_s and F_g of sand and gravel, respectively in the surface layer, the submerged specific gravity of the sediment R and shear velocity associated with skin friction u_*.

Kuhnle Et al.

For the case in which sand fraction is transported by the current over and through an immobile gravel bed, Kuhnle et al.(2013), following the theoretical analysis done by Pellachini (2011), provides a new relationship for the bed load transport of the sand fraction when gravel particles remain at rest. It is worth mentioning that Kuhnle et al.(2013) applied the Wilcock and Kenworthy (2002) formula to their experimental data and found out that predicted bed load rates of sand fraction were about 10 times greater than measured and approached 1 as the sand elevation became near the top of the gravel layer. They, also, hypothesized that the mismatch between predicted and measured sand bed load rates is due to the fact that the bed shear stress used for the Wilcock and Kenworthy (2002) formula was larger than that available for transport within the gravel bed because of the sheltering effect of the gravel particles. To overcome this mismatch, following Pellachini (2011), they assumed that the variability of the bed shear stress available for the sand to be transported by the current would be some function of the so-called "Roughness Geometry Function" (RGF), which represents the gravel bed elevations distribution. Therefore, the sand bed load formula follows as:

$$q_s^* = 2.29*10^{-5} A(z_s)^{2.14} \left(\frac{\tau_b}{\tau_{cs}} \right)^{3.49}$$

where

$$q_s^* = \frac{q_s}{[(s-1)gD_s]^{0.5} \rho_s D_s}$$

the subscript $_s$ refers to the sand fraction, s represents the ratio ρ_s / ρ_w where ρ_s is the sand fraction density, $A(z_s)$ is the RGF as a function of the sand level z_s within the gravel bed, τ_b is the bed shear stress available for sand transport and τ_{cs} is the critical shear stress for incipient motion of the sand fraction, which was calculated graphically using the updated Shields-type relation of Miller et al.(1977).

Suspended Load

Suspended load is carried in the lower to middle parts of the flow, and moves at a large fraction of the mean flow velocity in the stream.

A common characterization of suspended sediment concentration in a flow is given by the Rouse Profile. This characterization works for the situation in which sediment concentration c_0 at one particular elevation above the bed z_0 can be quantified. It is given by the expression:

$$\frac{c_s}{c_0} = \left[\frac{z(h-z_0)}{z_0(h-z)} \right]^{-P/\alpha}$$

Here, z is the elevation above the bed, c_s is the concentration of suspended sediment at that elevation, h is the flow depth, P is the Rouse number, and α relates the eddy viscosity for momentum K_m to the eddy diffusivity for sediment, which is approximately equal to one.

$$\alpha = \frac{K_s}{K_m} \approx 1$$

Experimental work has shown that α ranges from 0.93 to 1.10 for sands and silts.

The Rouse profile characterizes sediment concentrations because the Rouse number includes both turbulent mixing and settling under the weight of the particles. Turbulent mixing results in the net motion of particles from regions of high concentrations to low concentrations. Because particles settle downward, for all cases where the particles are not neutrally buoyant or sufficiently light that this settling velocity is negligible, there is a net negative concentration gradient as one goes upward in the flow. The Rouse Profile therefore gives the concentration profile that provides a balance between turbulent mixing (net upwards) of sediment and the downwards settling velocity of each particle.

Bed Material Load

Bed material load comprises the bed load and the portion of the suspended load that is sourced from the bed.

Three common bed material transport relations are the "Ackers-White", "Engelund-Hansen", "Yang" formulae. The first is for sand to granule-size gravel, and the second and third are for sand though Yang later expanded his formula to include fine gravel. That all of these formulae cover the sand-size range and two of them are exclusively for sand is that the sediment in sand-bed rivers is commonly moved simultaneously as bed and suspended load.

Engelund-Hansen

The bed material load formula of Engelund and Hansen is the only one to not include some kind of critical value for the initiation of sediment transport. It reads:

$$q_s* = \frac{0.05}{c_f} \tau *^{2.5}$$

where q_s* is the Einstein nondimensionalization for sediment volumetric discharge per unit width, c_f is a friction factor, and $\tau*$ is the Shields stress. The Engelund-Hansen formula is one of the few sediment transport formulae in which a threshold "critical shear stress" is absent.

Wash Load

Wash load is carried within the water column as part of the flow, and therefore moves with the mean velocity of main stream. Wash load concentrations are approximately uniform in the water column. This is described by the endmember case in which the Rouse number is equal to 0 (i.e. the settling velocity is far less than the turbulent mixing velocity), which leads to a prediction of a perfectly uniform vertical concentration profile of material.

Total Load

Some authors have attempted formulations for the total sediment load carried in water. These formulas are designed largely for sand, as (depending on flow conditions) sand often can be carried as both bed load and suspended load in the same stream or shoreface.

Rain Gauge

A rain gauge (also known as an udometer, pluviometer, or an ombrometer) is a type of instrument used by meteorologists and hydrologists to gather and measure the amount of liquid precipitation over a set period of time.

History

The first known rainfall records were kept by the Ancient Greeks, about 500 B.C.About 400 B.C. people in India also began to record rainfall. The readings were correlated against expected growth, and used as a basis for land taxes. In the Arthashastra, used for example in Magadha, precise standards were set as to grain production. Each of the state storehouses were equipped with a rain gauge to classify land for taxation purposes.

In 1441, the Cheugugi was invented during the reign of King Sejong the Great of the Joseon Dynasty in Korea as the first standardized rain gauge. In 1662, Christopher Wren created the first tipping-bucket rain gauge in Britain in collaboration with Robert Hooke. Hooke also designed a manual gauge with a funnel that made measurements throughout 1695.

It was Richard Towneley who was the first to make systematic rainfall measurements over a period of 15 years from 1677 to 1694, publishing his records in the *Philosophical Transactions of the Royal Society*. Towneley called for more measurements elsewhere in the country to compare the rainfall in different regions, although only William Derham appears to have taken up Towneley's challenge. They jointly published the rainfall measurements for Towneley Park and Upminster in Essex for the years 1697 to 1704.

The naturalist Gilbert White took measurements to determine the mean rainfall from 1779 to 1786, although it was his brother in law, Thomas Barker who made regular and meticulous measurements for 59 years, recording temperature, wind, barometric pressure, rainfall and clouds. His meteorological records are a valuable resource for knowledge of the 18th century British climate. He was able to demonstrate that the average rainfall varied greatly from year to year with little discernible pattern.

National Coverage and Modern Gauges

The meteorologist George James Symons published the first annual volume of *British Rainfall* in 1860. This pioneering work contained rainfall records from 168 land stations in England and Wales. He was elected to the council of the British meteorological society in 1863 and made it his life's work to investigate rainfall within the British Isles. He set up a voluntary network of observers, who collected data which was returned to him for analysis. So successful was he in this object that by 1866 he was able to show results which gave a fair representation of the distribution of rainfall, and the number of recorders gradually increased until the last volume of British Rainfall which he lived to edit (that for 1899) contained figures from 3528 stations — 2894 in England and Wales, 446 in Scotland, and 188 in Ireland. He also collected old rain fall records going back over a hundred years. In 1870 he produced an account of rainfall in the British Isles starting in 1725.

George James Symons, founder of the first systematic rainfall survey on a national basis.

Due to the ever increasing numbers of observers, standardisation of the gauges became necessary. Symons began experimenting on new gauges in his own garden. He tried different models with variations in size, shape, and height. In 1863 he began collaboration with Colonel Michael Foster Ward from Calne, Wiltshire, who undertook more extensive investigations. By including Ward and various others around Britain, the investigations continued until 1890. The experiments were remarkable for their planning, execution, and drawing of conclusions. The results of these experiments led to the progressive adoption of the well known standard gauge, still used by the UK Meteorological Office today. Namely, one made of '...copper, with a five-inch funnel having its brass rim one foot above the ground...'

Most rain gauges generally measure the precipitation in millimetres equivalent to litres per square metre. The level of rainfall is sometimes reported as inches or centimetres.

Rain gauge amounts are read either manually or by automatic weather station (AWS). The frequency of readings will depend on the requirements of the collection agency. Some countries will supplement the paid weather observer with a network of volunteers to obtain precipitation data (and other types of weather) for sparsely populated areas.

In most cases the precipitation is not retained, however some stations do submit rainfall (and snowfall) for testing, which is done to obtain levels of pollutants.

Rain gauges have their limitations. Attempting to collect rain data in a hurricane can be nearly impossible and unreliable (even if the equipment survives) due to wind extremes. Also, rain gauges only indicate rainfall in a localized area. For virtually any gauge, drops will stick to the sides or funnel of the collecting device, such that amounts are very slightly underestimated, and those of .01 inches or .25 mm may be recorded as a trace.

Another problem encountered is when the temperature is close to or below freezing. Rain may fall on the funnel and ice or snow may collect in the gauge, blocking subsequent rain.

Rain gauges should be placed in an open area where there are no obstacles, such as buildings or trees, to block the rain. This is also to prevent the water collected on the roofs of buildings or the leaves of trees from dripping into the rain gauge after a rain, resulting in inaccurate readings.

Types

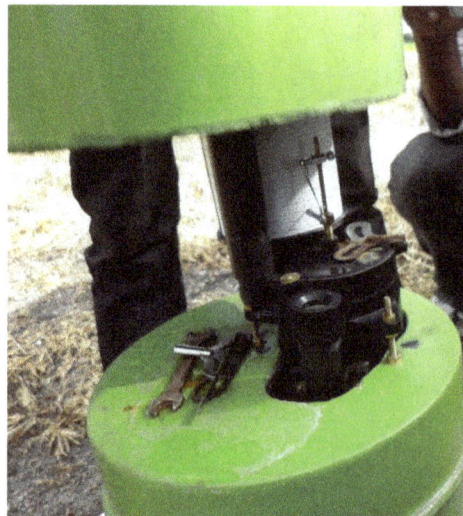

A self-recording rain gauge (interior)

Types of rain gauges include graduated cylinders, weighing gauges, tipping bucket gauges, and simple buried pit collectors. Each type has its advantages and disadvantages for collecting rain data.

Standard Rain Gauge

The standard NWS rain gauge, developed at the start of the 20th century, consists of a funnel emptying into a graduated cylinder, 2 cm in radius, which fits inside a larger container which is 20 cm in diameter and 50 cm tall. If the rainwater overflows the graduated inner cylinder, the larger outer container will catch it. When measurements are taken, the height of the water in the small graduated cylinder is measured, and the excess overflow in the large container is carefully poured into another graduated cylinder and measured to give the total rainfall. Sometimes a cone meter is used to prevent leakage that can result in alteration of the data. In locations using the metric system, the cylinder is usually marked in mm and will measure up to 250 millimetres (9.8 in) of rainfall. Each horizontal line on the cylinder is 0.5 millimetres (0.02 in). In areas using Imperial units each horizontal line represents 0.01 inch.

Pluviometer of Intensities

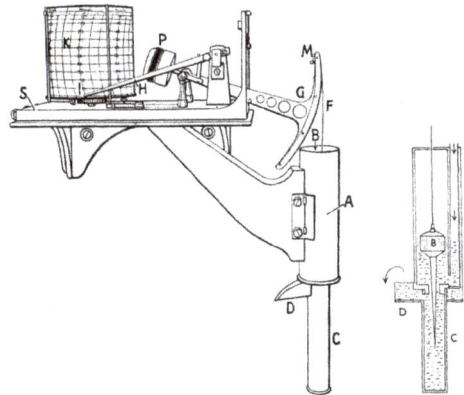

Pluviometer of intensities (1921)

The pluviometer of intensities (or Jardi's pluviometer) is a tool that measures the average intensity of rainfall in a certain interval of time. It was initially designed to record the rainfall regime in Catalonia but eventually spread globally throughout the world.

It employs the principle of feedback .. the incoming water pushes the buoy upwards , making the lower "adjusting conic needle" to let pass the same amount of water that enters into the container, this way.. the needle records on the drum the amount of water flowing through it at every moment -in mm of rainfall per square-meter

It consists of a rotating drum that rotates at constant speed , this drum drags a graduate sheet of cardboard, which has the time at the abscissa while the y axis indicates the height of rainfall in mm of rain. This height is recorded with a pen that moves vertically, driven by a buoy, marking on the paper the rainfall over the time (the cardboard sheet is usually for one day).

While the rain falls, the water collected by the funnel falls into the container and raises the buoy ... , that makes the pen arm raising in the vertical axis marking the cardboard accordingly. If the rainfall does not vary, the water level in the container remains constant, and while the drum rotates, the pen's mark it's more or less a horizontal line, proportional to the amount of water that is falling. When the pen reaches the top edge of the recording paper, it means that the buoy is "up high in the tank" leaving the tip of the conic needle in a way that uncovers the regulating hole (i.e. the maximum flow that the apparatus is able to record) . If the rain suddenly decreases, making the container -as it empties- to quickly lower the buoy, that movement corresponds to a steep slope line that can reach the bottom of the recorded cardboard, if it stops raining.

The pluviometer of intensities allowed to record the precipitation over the time and the years, particularly in Barcelona (95 years), apart from many other places around the world, such as Hong-Kong.

Weighing Precipitation Gauge

A weighing-type precipitation gauge consists of a storage bin, which is weighed to record the mass. Certain models measure the mass using a pen on a rotating drum, or by using a vibrating wire attached to a data logger. The advantages of this type of gauge over tipping buckets are that it does not underestimate intense rain, and it can measure other forms of precipitation, including rain, hail and snow. These gauges are, however, more expensive and require more maintenance than tipping bucket gauges.

The weighing-type recording gauge may also contain a device to measure the quantity of chemicals contained in the location's atmosphere. This is extremely helpful for scientists studying the effects of greenhouse gases released into the atmosphere and their effects on the levels of the acid rain. Some Automated Surface Observing System (ASOS) units use an automated weighing gauge called the AWPAG (All Weather Precipitation Accumulation Gauge).

Tipping Bucket Rain Gauge

The interior of a tipping bucket rain gauge

The tipping bucket rain gauge consists of a funnel that collects and channels the precipitation into a small seesaw-like container. After a pre-set amount of precipitation falls, the lever tips, dumping the collected water and sending an electrical signal. An old-style recording device may consist of a pen mounted on an arm attached to a geared wheel that moves once with each signal sent from the collector. In this design, the wheel turns the pen arm moves either up or down leaving a trace on the graph and at the same time making a loud click. Each jump of the arm is sometimes referred to as a 'click' in reference to the noise. The chart is measured in 10-minute periods (vertical lines) and 0.4 mm (0.015 in) (horizontal lines) and rotates once every 24 hours and is powered by a clockwork motor that must be manually wound.

The exterior of a tipping bucket rain gauge

The tipping bucket rain gauge is not as accurate as the standard rain gauge because the rainfall may stop before the lever has tipped. When the next period of rain begins it may take no more than one or two drops to tip the lever. This would then indicate that pre-set amount has fallen when in fact only a fraction of that amount has actually fallen. Tipping buckets also tend to underestimate the amount of rainfall, particularly in snowfall and heavy rainfall events. The advantage of the tipping bucket rain gauge is that the character of the rain (light, medium, or heavy) may be easily obtained. Rainfall character is decided by the total amount of rain that has fallen in a set period (usually 1 hour) and by counting the number of 'clicks' in a 10-minute period the observer can decide the character of the rain. Correction algorithms can be applied to the data as an accepted method of correcting the data for high level rainfall intensity amounts.

Tipping bucket rain gauge recorder

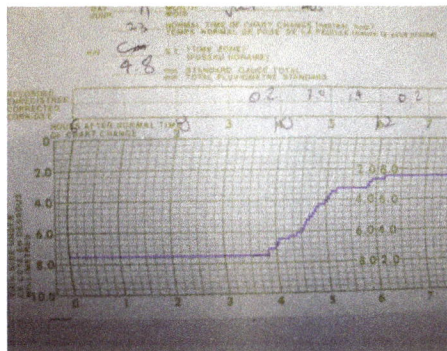

Close up of a tipping bucket rain gauge recorder chart

Modern tipping rain gauges consist of a plastic collector balanced over a pivot. When it tips, it actuates a switch (such as a reed switch) which is then electronically recorded or transmitted to a remote collection station.

Tipping gauges can also incorporate weighing gauges. In these gauges, a strain gauge is fixed to the collection bucket so that the exact rainfall can be read at any moment. Each time the collector tips, the strain gauge (weight sensor) is re-zeroed to null out any drift.

To measure the *water equivalent* of frozen precipitation, a tipping bucket may be heated to melt any ice and snow that is caught in its funnel. Without a heating mechanism, the funnel often becomes clogged during a frozen precipitation event, and thus no precipitation can be measured. Many Automated Surface Observing System (ASOS) units use heated tipping buckets to measure precipitation

Optical Rain Gauge

These have a row of collection funnels. In an enclosed space below each is a laser diode and a photo transistor detector. When enough water is collected to make a single drop, it drops from the bottom, falling into the laser beam path. The sensor is set at right angles to the laser so that enough light is scattered to be detected as a sudden flash of light. The flashes from these photo detectors are then read and transmitted or recorded.

Acoustic Rain Gauge

The acoustic disdrometer developed by Stijn de Jong is an acoustic rain gauge. Also referred to as a hydrophone, it is able to sense the sound signatures for each drop size as rain strikes a water surface within the gauge. Since each sound signature is unique, it is possible to invert the underwater sound field to estimate the drop-size distribution within the rain. Selected moments of the drop-size distribution yield rainfall rate, rainfall accumulation, and other rainfall properties.

Stream Gauge

Brant Broughton Gauging Station on the River Brant in Lincolnshire, England.

A stream gauge, streamgage or gauging station is a location used by hydrologists or environmental scientists to monitor and test terrestrial bodies of water. Hydrometric measurements of water level surface elevation ("stage") and/or volumetric discharge (flow) are generally taken and observations of biota and water quality may also be made. The location of gauging stations are often found on topographical maps. Some gauging stations are highly automated and may include telemetry capability transmitted to a central data logging facility.

Automated direct measurement of streamflow discharge is difficult at present. In place of the direct measurement of streamflow discharge, one or more surrogate measurements can be used to produce discharge values. In the majority of cases, a stage (the elevation of the water surface) measurement is used as the surrogate. Low gradient (or shallow-sloped) streams are highly influenced by variable downstream channel conditions. For these streams, a second stream gauge would be installed, and the slope of the water surface would be calculated between the gauges. This value would be used along with the stage measurement to more accurately determine the streamflow discharge. Within the last ten years, the technological advance of velocity sensors has allowed the use of water velocity as a reliable surrogate for streamflow discharge at sites with a stable cross-sectional area. These sensors are permanently mounted in the stream and measure velocity at a particular location in the stream and related to flow in a manner similar to the use of traditional water level.

Stream Gaging Station, Carnation, Washington data available at .

In those instances where only a stage measurement is used as the surrogate, a rating curve must be constructed. A rating curve is the functional relation between stage and discharge. It is determined by making repeated discrete measurements of streamflow discharge using a velocimeter and some means to measure the channel geometry to determine the cross-sectional area of the channel. The technicians and hydrologists responsible for determining the rating curve visit the site routinely, with special trips to measure the hydrologic extremes (floods and droughts), and make a discharge measurement by following an explicit set of instructions.

December 12, 2001 photo of the USGS streamflow-gaging station at Huey Creek, McMurdo Dry Valleys, Antarctica.

Once the rating curve is established, it can be used in conjunction with stage measurements to determine the volumetric streamflow discharge. This record then serves as an assessment of the volume of water that passes by the stream gauge and is useful for many tasks associated with hydrology.

In those instances where a velocity measurement is additionally used as a surrogate, an index velocity determination is conducted. This analysis uses a velocity sensor, often either magnetic or acoustic, to measure the velocity of the flow at a particular location in the stream cross section. Once again, discrete measurements of streamflow discharge are made by the technician or hydrologist at a variety of stages. For each discrete determination of streamflow discharge, the mean velocity of the cross section is determined by dividing streamflow discharge by the cross-sectional area. A rating curve, similar to that used for stage-discharge determinations, is constructed using the mean velocity and the index velocity from the permanently mounted meter. An additional rating curve is constructed that relates stage of the stream to cross-sectional area. Using these two ratings, the automatically collected stage produces an estimate of the cross-sectional area, and the automatically collected index velocity produces an estimate of the mean velocity of the cross section. The streamflow discharge is computed as the estimate of the cross section area and the estimate of the mean velocity of the streamflow.

Stream gauge B62, a combination weir at Doddieburn, on the Mzingwane River, Zimbabwe

A variety of hydraulic structures / primary device are used to improve the reliability of using water level as a surrogate for flow (improving the accuracy of the rating table), including:

- Weirs

  o V-notch,

  o broad-crested,

  o sharp-crested and

  o combination weirs

- Flumes

  o Parshall flume

Other equipment commonly used at permanent stream gauge include:

- Cableways - for suspending a hydrographer and current meter over a river to make high flow measurement

- Stilling well - to provide a calm water level that can be measured by a sensor

Water level gauges:

- Staff (head) gauges - for a visual indication of water depth

- Water pressure measuring device (Bubbler) - to measure water level via pressure (typically done directly in-stream without a stilling well)

- Stage encoder - a potentiometer with a wheel and pulley system connected to a float in a stilling well to provide an electronic reading of the water level

- Simple ultrasonic devices - to measure water level in a stilling well or directly in a canal.

- Electromagnetic gauges

Discharge measurements of a stream or canal without an established stream gage can be made using a current meter or Acoustic Doppler current profiler. One informal methods that is not acceptable for any official or scientific purpose, but can be useful is the float method, in which a floating object such as a piece of wood or orange peel is observed floating down the stream.

National Stream Gauge Networks

Plaque marking the construction of the River Dove gauging station, dedicated to Izaak Walton, author of The Compleat Angler.

United Kingdom

The first routine measurements of river flow in England began on the Thames and Lea in the 1880s, and in Scotland on the River Garry in 1913. The national gauging station network was established in its current form by the early 1970s and consists of approximately 1500 flow measurement stations supplemented by a variable number of temporary monitoring sites. The Environment Agency is responsible for collection and analysis of hydrometric data in England and Wales, whilst responsibility for Scotland and Northern Ireland rests with the Scottish Environment Protection Agency and Rivers Agency respectively.

United States

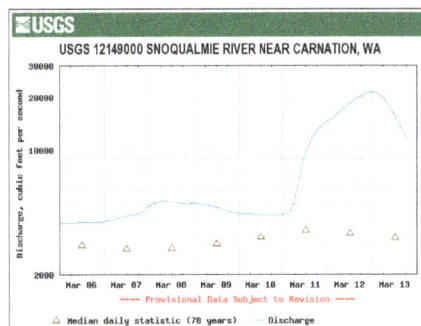

Hydrograph of the March 13, 2007 Snoqualmie River flood

In the United States, the U.S. Geological Survey (USGS) is the principal federal agency tasked with maintaining records of natural resources. Within the USGS, the Water Resources Division carries the responsibility for monitoring water resources.

To establish a stream gauge, USGS personnel first choose a site on a stream where the geometry is relatively stable and there is a suitable location to make discrete direct measurements of streamflow using specialized equipment. Many times this will be at a bridge or other stream crossing. Technicians then install equipment that measures the stage (the elevation of the water surface) or, more rarely, the velocity of the flow. Additional equipment is installed to record and transmit these readings (via the Geostationary Operational Environmental Satellite) to the Water Science Center office where the records are kept. The USGS has a Water Science Center office in every state within the United States. Current streamflow data from USGS streamgages may be viewed in map form at: .

Zimbabwe

In Zimbabwe, the national stream gauge network is the responsibility of the Zimbabwe National Water Authority. This is an extensive network covering all major rivers and catchments in the country. However, a review of existing gauges raised serious concerns about the reliability of the data of a minority of stations, due in part to ongoing funding problems.

Czech Republic

Emergency levels I, II and III at Tichá Orlice river in Choceň, Czech Republic

In the Czech Republic, in some measuring places (profiles) are defined three levels which define three degrees of flood-emergency activity. The degree I is a situation of alertness, the degree II is a situation of readiness, the degree III is a situation of danger.

Hygrometer

A hygrometer is an instrument used for measuring the moisture content in the atmosphere. Humidity measurement instruments usually rely on measurements of some other quantity such as temperature, pressure, mass or a mechanical or electrical change in a substance as moisture is absorbed. By calibration and calculation, these measured quantities can lead to a measurement of humidity. Modern electronic devices use temperature of condensation (the dew point), or changes in electrical capacitance or resistance to measure humidity differences. The first crude hygrometer was invented by Leonardo da Vinci in 1480 and a more modern version was created by polymath Johann Heinrich Lambert in 1755.

A Haar tension dial hygrometer with a nonlinear scale.

Classical Hygrometers

Metal-paper Coil Type

The metal-paper coil hygrometer is very useful for giving a dial indication of humidity changes. It appears most often in very inexpensive devices, and its accuracy is limited, with variations of 10% or more. In these devices, water vapour is absorbed by a salt-impregnated paper strip attached to a metal coil, causing the coil to change shape. These changes (analogous to those in a bimetallic thermometer) cause an indication on a dial.

Hair Tension Hygrometers

These devices use a human or animal hair under tension. The hair is hygroscopic (tending toward retaining moisture); its length changes with humidity, and the length change may be magnified by a mechanism and indicated on a dial or scale. In the late 1700s, such devices were called by some scientists *hygroscopes*; that word is no longer in current use, but *hygroscopic* and *hygroscopy*, which derive from it, still are. The traditional folk art device known as a weather house works on this principle. Whale bone and other materials may be used in place of hair.

In 1783, Swiss physicist and geologist Horace Bénédict de Saussure built the first hair-tension hygrometer using human hair.

It consists of a human hair eight to ten inches long, b c, fastened at one extremity to a screw, a, and at the other passing over a pulley, c, being strained tight by a silk thread and weight, d.

—John William Draper, A Textbook on Chemistry

The pulley is connected to an index which moves over a graduated scale (e). The instrument can be made more sensitive by removing oils from the hair, such as by first soaking the hair in diethyl ether.

Psychrometer (Wet-and-dry-bulb thermometer)

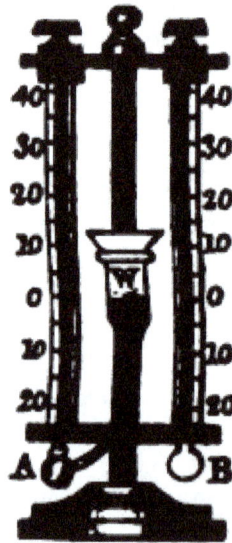

1861 diagram of a psychrometer with wet bulb (a) and dry bulb (b). The wet bulb is connected to a reservoir of water.

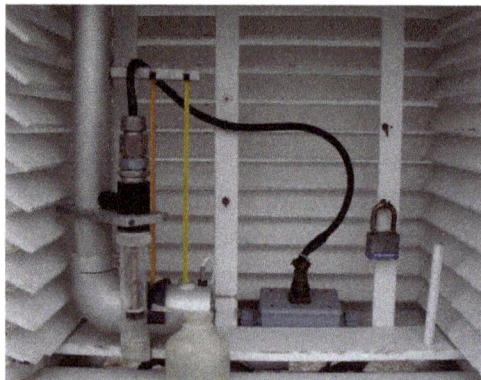

The interior of a Stevenson screen showing a motorized psychrometer

A psychrometer, or wet-and-dry-bulb thermometer, consists of two thermometers, one that is dry and one that is kept moist with distilled water on a sock or wick. At temperatures above the freezing point of water, evaporation of water from the wick lowers the temperature, so that the wet-bulb thermometer usually shows a lower temperature than that of the dry-bulb thermometer. When the air temperature is below freezing, however, the wet-bulb is covered with a thin coating of ice and may be warmer than the dry bulb.

Relative humidity is computed from the ambient temperature as shown by the dry-bulb thermometer and the difference in temperatures as shown by the wet-bulb and dry-bulb thermometers. Relative humidity can also be determined by locating the intersection of the wet and dry-bulb temperatures on a psychrometric chart. The two thermometers coincide when the air is fully saturated, and the greater the difference the drier the air. Psychrometers are commonly used in meteorology, and in the HVAC industry for proper refrigerant charging of residential and commercial air conditioning systems.

Sling Psychrometer

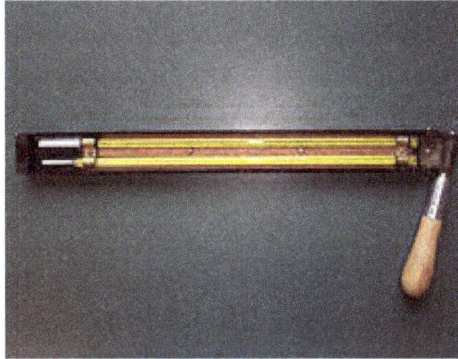

A sling psychrometer for outdoor use

A sling psychrometer, which uses thermometers attached to a handle or length of rope and spun in the air, is sometimes used for field measurements, but is being replaced by more convenient electronic sensors. A whirling psychrometer uses the same principle, but the two thermometers are fitted into a device that resembles a ratchet or football rattle.

Electronic hygrometer

Chilled Mirror Dew Point Hygrometer

Dew point is the temperature at which a sample of moist air (or any other water vapor) at constant pressure reaches water vapor saturation. At this saturation temperature, further cooling results in condensation of water. Chilled mirror dewpoint hygrometers are some of the most precise instruments commonly available. They use a chilled mirror and optoelectronic mechanism to detect condensation on the mirror's surface. The temperature of the mirror is controlled by electronic feedback to maintain a dynamic equilibrium between evaporation and condensation, thus closely measuring the dew point temperature. An accuracy of 0.2 °C is attainable with these devices, which correlates at typical office environments to a relative humidity accuracy of about ±1.2%. These devices need frequent cleaning, a skilled operator and periodic calibration to attain these levels of accuracy. Even so, they are prone to heavy drifting in environments where smoke or otherwise impure air may be present.

Modern Hygrometers

Capacitive

For applications where cost, space, or fragility are relevant, other types of electronic sensors are used, at the price of a lower accuracy. In capacitive hygrometers, the effect of humidity on the dielectric constant of a polymer or metal oxide material is measured. With calibration, these sensors have an accuracy of ±2% RH in the range 5–95% RH. Without calibration, the accuracy is 2 to 3 times worse. Capacitive sensors are robust against effects such as condensation and temporary high temperatures. Capacitive sensors are subject to contamination, drift and aging effects, but are suitable for many applications.

Resistive

In resistive hygrometers, the change in electrical resistance of a material due to humidity is measured. Typical materials are salts and conductive polymers. Resistive sensors are less sensitive than capacitive sensors – the change in material properties is less, so they require more complex circuitry. The material properties also tend to depend both on humidity and temperature, which means in practice that the sensor must be combined with a temperature sensor. The accuracy and robustness against condensation vary depending on the chosen resistive material. Robust, condensation-resistant sensors exist with an accuracy of up to ±3% RH.

Thermal

In thermal hygrometers, the change in thermal conductivity of air due to humidity is measured. These sensors measure absolute humidity rather than relative humidity.

Gravimetric

A gravimetric hygrometer measures the mass of an air sample compared to an equal volume of dry air. This is considered the most accurate primary method to determine the moisture content of the air. National standards based on this type of measurement have been developed in US, UK, EU and Japan. The inconvenience of using this device means that it is usually only used to calibrate less accurate instruments, called transfer standards.

Applications

Aside from greenhouses and industrial spaces, hygrometers are also used in some incubators, saunas, humidors and museums. They are also used in the care of wooden musical instruments such as pianos, guitars, violins, and harps which can be damaged by improper humidity conditions. In residential settings, hygrometers are used to assist in humidity control (too low humidity can damage human skin and body, while too high humidity favors growth of mildew and dust mite). Hygrometers are also used in the coating industry because the application of paint and other coatings may be very sensitive to humidity and dew point. With a growing demand on the amount of measurements taken the psychrometer is now replaced by a dewpoint gauge known as a dewcheck. These devices make measurements a lot faster but are often not allowed in explosive environments.

Difficulty of Accurate Humidity Measurement

Humidity measurement is among the more difficult problems in basic meteorology. According to the WMO Guide, "The achievable accuracies [for humidity determination] listed in the table refer to good quality instruments that are well operated and maintained. In practice, these are not easy to achieve." Two thermometers can be compared by immersing them both in an insulated vessel of water (or alcohol, for temperatures below the freezing point of water) and stirring vigorously to minimize temperature variations. A high-quality liquid-in-glass thermometer if handled with care should remain stable for some years. Hygrometers must be calibrated in air, which is a much less effective heat transfer medium than is water, and many types are subject to drift so need regular recalibration. A further difficulty is that most hygrometers sense relative humidity rather than the absolute amount of water present, but relative humidity is a function of both temperature and absolute moisture content, so small temperature variations within the air in a test chamber will translate into relative humidity variations.

In a cold and humid environment, sublimation of ice may occur on the sensor head, whether it is a hair, dew cell, mirror, capacitance sensing element, or dry-bulb thermometer of an aspiration psychrometer. The ice on the probe matches the reading to the saturation humidity with respect to ice at that temperature, i.e. the frost point. However, a conventional hygrometer is unable to measure properly above the frost point, and the only way to go around this fundamental problem is to use a heated humidity probe.

Calibration Standards

Psychrometer Calibration

Accurate calibration of the thermometers used is fundamental to precise humidity determination by the wet-dry method. The thermometers must be protected from radiant heat and must have a sufficiently high flow of air over the wet bulb for the most accurate results. One of the most precise types of wet-dry bulb psychrometer was invented in the late 19th century by Adolph Richard Aßmann (1845–1918); in English-language references the device is usually spelled "Assmann psychrometer." In this device, each thermometer is suspended within a vertical tube of polished metal, and that tube is in turn suspended within a second metal tube of slightly larger diameter; these double tubes serve to isolate the thermometers from radiant heating. Air is drawn through the tubes with a fan that is driven by a clockwork mechanism to ensure a consistent speed (some modern versions use an electric fan with electronic speed control). According to Middleton, 1966, "an essential point is that air is drawn between the concentric tubes, as well as through the inner one."

It is very challenging, particularly at low relative humidity, to obtain the maximal theoretical depression of the wet-bulb temperature; an Australian study in the late 1990s found that liquid-in-glass wet-bulb thermometers were warmer than theory predicted even when considerable precautions were taken; these could lead to RH value readings that are 2 to 5 percent points too high.

One solution sometimes used for accurate humidity measurement when the air temperature is below freezing is to use a thermostatically-controlled electric heater to raise the temperature of outside air to above freezing. In this arrangement, a fan draws outside air past (1) a thermometer to measure the ambient dry-bulb temperature, (2) the heating element, (3) a second thermometer to measure the dry-bulb temperature of the heated air, then finally (4) a wet-bulb thermometer. According to the World Meteorological Organization Guide, "The principle of the heated psychrometer is that the water vapour content of an air mass does not change if it is heated. This property may be exploited to the advantage of the psychrometer by avoiding the need to maintain an ice bulb under freezing conditions.".

Since the humidity of the ambient air is calculated indirectly from three temperature measurements, in such a device accurate thermometer calibration is even more important than for a two-bulb configuration.

Saturated Salt Calibration

Various researchers have investigated the use of saturated salt solutions for calibrating hygrometers. Slushy mixtures of certain pure salts and distilled water have the property that they maintain an approximately constant humidity in a closed container. A saturated table salt (Sodium Chloride) bath will eventually give a reading of approximately 75%. Other salts have other equilibrium humidity levels: Lithium Chloride ~11%; Magnesium Chloride ~33%; Potassium Carbonate ~43%; Potassium Sulfate ~97%. Salt solutions will vary somewhat in humidity with temperature and they can take relatively long times to come to equilibrium, but their ease of use compensates somewhat for these disadvantages in low precision applications, such as checking mechanical and electronic hygrometers.

Green Kenue

Green Kenue (formerly EnSim Hydrologic) is an advanced data preparation, analysis, and visualization tool for hydrologic modellers. It is a Windows/OpenGL-based graphical user interface, integrating environmental databases and geo-spatial data with model input and results data. Green Kenue provides complete pre- and post-processing for the WATFLOOD and HBV-EC hydrologic models. Also included is a 1D "reach scale" unsteady hydrodynamic flow solver, Gen1D.

Visualization and Animation

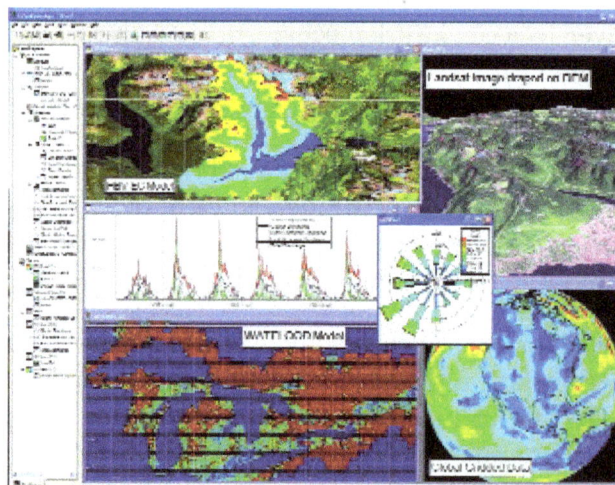

Visualization in Green Kenue is provided by dynamic 1D, Polar, 2D, 3D and Spherical views that can be recorded as digital movies or saved as images for inclusion in reports or presentations.

All views and data are fully geo-referenced and coordinate conversion between common projections is supported.

Data Formats/Types

Green Kenue provides support for a full-featured set of data types commonly used by hydrologic modellers. ASCII and Binary native file formats (for static and temporally-varying data) are available.

- Time-series (scalar and vector)

- Tabular data, Distributions, Velocity roses

- Multi-attribute point-sets

- Multi-attribute line-sets

- Multi-attribute networks

- Rectangular gridded data (scalar and vector)

- Triangular gridded data (scalar and vector)

Green Kenue supports import and export for common GIS data formats including:

- ArcINFO/ArcView,

- MapInfo,

- GeoTIFF,

- DTED/CDED,

- BIL,

- SRTM,

- WMO-GRIB,

- Surfer

Digital Elevation Map Processing

The watershed basin and stream delineation tool uses either the Jenson or At Search algorithm and allows for artificial diversions. A suite of sophisticated basin metrics and digital elevation map analysis tools is provided, including statistical and analysis functions such as temporal min, max, mean, standard deviation, slope and curvature analysis, hypsography, stream power, and relief potential. For more sophisticated analysis, a powerful algebraic calculator for gridded data is available.

Environmental Databases

Green Kenue provides an interface to Environment Canada's hydrometric station database (HYDAT) as well as the Canadian Daily Climate Database (CDCD). Stations are queried interactively and available time series data can be extracted, re-sampled, analyzed, and processed with various editors and calculators. Users can also load and analyze gridded weather forecast datasets in GRIB or FST format.

References

- King, compiled by James J. (1995). The environmental dictionary : and regulatory cross reference (3rd ed.). New York: Wiley. p. 121. ISBN 0-471-11995-4.

- Atif Ansar; Bent Flyvbjerg; Alexander Budzier; Daniel Lunn (June 2014). "Should we build more large dams? The actual costs of hydropower megaproject development". Energy Policy, vol. 69, pp. 43-56.

- "Lake Diefenbaker Reservoir Operations Context and Objectives" (PDF). Saskatchewan Watershed Authority. Retrieved 27 June 2013.

- "Dams the latest culprit in global warming". Times of India. 8 August 2012. Archived from the original on 9 August 2012. Retrieved 9 August 2012.

- The long profile – changing processes: types of erosion, transportation and deposition, types of load; the Hjulstrom curve. coolgeography.co.uk. Last accessed 26 Dec 2011.

- Special Topics: An Introduction to Fluid Motions, Sediment Transport, and Current-generated Sedimentary Structures; As taught in: Fall 2006. Massachusetts Institute of Technology. 2006. Last accessed 26 Dec 2011.

8

Elements of Hydrological Study

The science of hydrology incorporates numerous elements that study the ways and properties of geological water circulation and how water is continuously replenished naturally. This chapter discusses water cycle, water resources, water table, drainage basin, groundwater flow, aquifer, hydraulic conductivity, evaporation and moisture recycling. There is a section on water pollution as well.

Water Cycle

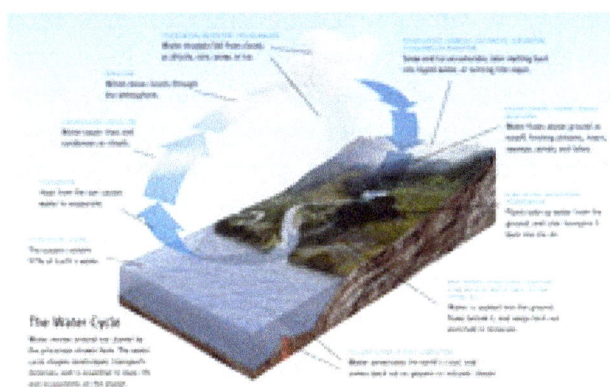

Diagram of the Water Cycle

The water cycle, also known as the hydrological cycle or the H_2O cycle, describes the continuous movement of water on, above and below the surface of the Earth. The mass of water on Earth remains fairly constant over time but the partitioning of the water into the major reservoirs of ice, fresh water, saline water and atmospheric water is variable depending on a wide range of climatic variables. The water moves from one reservoir to another, such as from river to ocean, or from the ocean to the atmosphere, by the physical processes of evaporation, condensation, precipitation, infiltration, surface runoff, and subsurface flow. In doing so, the water goes through different phases: liquid, solid (ice) and vapor.

The water cycle involves the exchange of energy, which leads to temperature changes. For instance, when water evaporates, it takes up energy from its surroundings and cools the environment. When it condenses, it releases energy and warms the environment. These heat exchanges influence climate.

The evaporative phase of the cycle purifies water which then replenishes the land with freshwater. The flow of liquid water and ice transports minerals across the globe. It is also involved in reshaping the geological features of the Earth, through processes including erosion and sedimentation. The water cycle is also essential for the maintenance of most life and ecosystems on the planet.

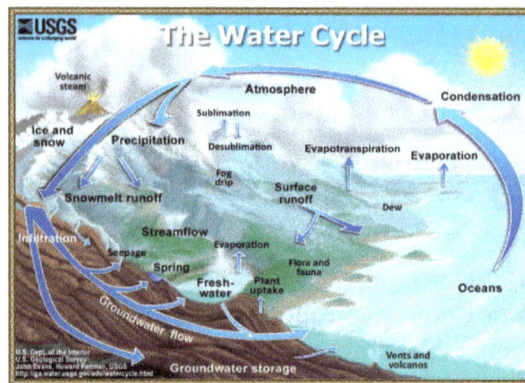

The water cycle

Description

The sun, which drives the water cycle, heats water in oceans and seas. Water evaporates as water vapour into the air. Ice, rain and snow can sublimate directly into water vapour. Evapotranspiration is
water transpired from plants and evaporated from the soil. Water vapour molecule H_2O, has less density compared to the major components of the atmosphere, nitrogen and oxygen, N_2 and O_2. Due to
the significant difference in molecular mass, water vapor in gas form gain height in open air as a result
of buoyancy. However, as altitude increases, air pressure decreases and the temperature drops. The
lowered temperature causes water vapour to condense into a tiny liquid water droplet which is heavier
than the air, such that it falls unless supported by an updraft. A huge concentration of these droplets
over a large space up in the atmosphere become visible as cloud. Fog is formed if the water vapour
condense near ground level, as a result of moist air and cool air collision or an abrupt reduction in air
pressure. Air currents move water vapour around the globe, cloud particles collide, grow, and fall out
of the upper atmospheric layers as precipitation. Some precipitation falls as snow or hail, sleet, and can
accumulate as ice caps and glaciers, which can store frozen water for thousands of years. Most water
falls back into the oceans or onto land as rain, where the water flows over the ground as surface runoff.
A portion of runoff enters rivers in valleys in the landscape, with streamflow moving water towards
the oceans. Runoff and water emerging from the ground (groundwater) may be stored as freshwater
in lakes. Not all runoff flows into rivers, much of it soaks into the ground as infiltration. Some water
infiltrates deep into the ground and replenishes aquifers, which can store freshwater for long periods of
time. Some infiltration stays close to the land surface and can seep back into surface-water bodies (and
the ocean) as groundwater discharge. Some groundwater finds openings in the land surface and comes
out as freshwater springs. In river valleys and flood-plains there is often continuous water exchange
between surface water and ground water in the hyporheic zone. Over time, the water returns to the
ocean, to continue the water cycle.

Processes

Precipitation

> Condensed water vapor that falls to the Earth's surface . Most precipitation occurs as rain,
> but also includes snow, hail, fog drip, graupel, and sleet. Approximately 505,000 km³
> (121,000 cu mi) of water falls as precipitation each year, 398,000 km³ (95,000 cu mi)
> of it over the oceans. The rain on land contains 107,000 km³ (26,000 cu mi) of water

per year and a snowing only 1,000 km³ (240 cu mi). 78% of global precipitation occurs over the ocean.

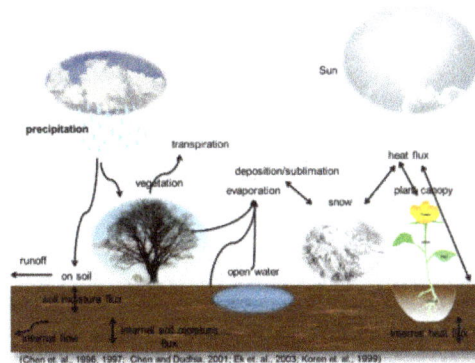

Many different processes lead to movements and phase changes in water

Canopy interception

The precipitation that is intercepted by plant foliage, eventually evaporates back to the atmosphere rather than falling to the ground.

Snowmelt

The runoff produced by melting snow.

Runoff

The variety of ways by which water moves across the land. This includes both surface runoff and channel runoff. As it flows, the water may seep into the ground, evaporate into the air, become stored in lakes or reservoirs, or be extracted for agricultural or other human uses.

Infiltration

The flow of water from the ground surface into the ground. Once infiltrated, the water becomes soil moisture or groundwater. A recent global study using water stable isotopes, however, shows that not all soil moisture is equally available for groundwater recharge or for plant transpiration.

Subsurface flow

The flow of water underground, in the vadose zone and aquifers. Subsurface water may return to the surface (e.g. as a spring or by being pumped) or eventually seep into the oceans. Water returns to the land surface at lower elevation than where it infiltrated, under the force of gravity or gravity induced pressures. Groundwater tends to move slowly, and is replenished slowly, so it can remain in aquifers for thousands of years.

Evaporation

The transformation of water from liquid to gas phases as it moves from the ground or bodies of water into the overlying atmosphere. The source of energy for evaporation is primarily solar radiation. Evaporation often implicitly includes transpiration from plants, though together they are specifically referred to as evapotranspiration. Total annual evapotranspiration amounts to approximately 505,000 km³ (121,000 cu mi) of water, 434,000 km³ (104,000 cu mi) of which evaporates from the oceans. 86% of global evaporation occurs over the ocean.

Sublimation

The state change directly from solid water (snow or ice) to water vapor.

Deposition

This refers to changing of water vapor directly to ice.

Advection

The movement of water — in solid, liquid, or vapor states — through the atmosphere. Without advection, water that evaporated over the oceans could not precipitate over land.

Condensation

The transformation of water vapor to liquid water droplets in the air, creating clouds and fog.

Transpiration

The release of water vapor from plants and soil into the air. Water vapor is a gas that cannot be seen.

Percolation

Water flows vertically through the soil and rocks under the influence of gravity

Plate tectonics

Water enters the mantle via subduction of oceanic crust. Water returns to the surface via volcanism.

Water cycle thus involves many of the intermediate processes.

Residence Times

Average reservoir residence times	
Reservoir	**Average residence time**
Antarctica	20,000 years
Oceans	3,200 years
Glaciers	20 to 100 years
Seasonal snow cover	2 to 6 months
Soil moisture	1 to 2 months

Groundwater: shallow	100 to 200 years
Groundwater: deep	10,000 years
Lakes	50 to 100 years
Rivers	2 to 6 months
Atmosphere	9 days

The *residence time* of a reservoir within the hydrologic cycle is the average time a water molecule will spend in that reservoir (*see adjacent table*). It is a measure of the average age of the water in that reservoir.

Groundwater can spend over 10,000 years beneath Earth's surface before leaving. Particularly old groundwater is called fossil water. Water stored in the soil remains there very briefly, because it is spread thinly across the Earth, and is readily lost by evaporation, transpiration, stream flow, or groundwater recharge. After evaporating, the residence time in the atmosphere is about 9 days before condensing and falling to the Earth as precipitation.

The major ice sheets - Antarctica and Greenland - store ice for very long periods. Ice from Antarctica has been reliably dated to 800,000 years before present, though the average residence time is shorter.

In hydrology, residence times can be estimated in two ways. The more common method relies on the principle of conservation of mass and assumes the amount of water in a given reservoir is roughly constant. With this method, residence times are estimated by dividing the volume of the reservoir by the rate by which water either enters or exits the reservoir. Conceptually, this is equivalent to timing how long it would take the reservoir to become filled from empty if no water were to leave (or how long it would take the reservoir to empty from full if no water were to enter).

An alternative method to estimate residence times, which is gaining in popularity for dating groundwater, is the use of isotopic techniques. This is done in the subfield of isotope hydrology.

Changes Over Time

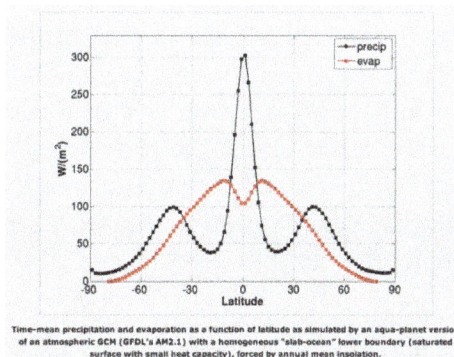

Time-mean precipitation and evaporation as a function of latitude as simulated by an aqua-planet version of an atmospheric GCM (GFDL's AM2.1) with a homogeneous "slab-ocean" lower boundary (saturated surface with small heat capacity), forced by annual mean insolation.

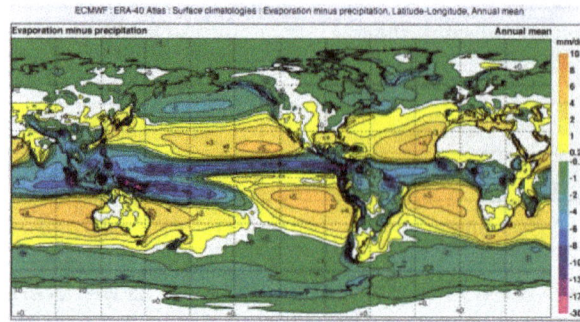

Global map of annual mean evaporation minus precipitation by latitude-longitude

The water cycle describes the processes that drive the movement of water throughout the hydrosphere. However, much more water is "in storage" for long periods of time than is actually moving through the cycle. The storehouses for the vast majority of all water on Earth are the oceans. It is estimated that of the 332,500,000 mi³ (1,386,000,000 km³) of the world's water supply, about 321,000,000 mi³ (1,338,000,000 km³) is stored in oceans, or about 97%. It is also estimated that the oceans supply about 90% of the evaporated water that goes into the water cycle.

During colder climatic periods more ice caps and glaciers form, and enough of the global water supply accumulates as ice to lessen the amounts in other parts of the water cycle. The reverse is true during warm periods. During the last ice age glaciers covered almost one-third of Earth's land mass, with the result being that the oceans were about 400 ft (122 m) lower than today. During the last global "warm spell," about 125,000 years ago, the seas were about 18 ft (5.5 m) higher than they are now. About three million years ago the oceans could have been up to 165 ft (50 m) higher.

The scientific consensus expressed in the 2007 Intergovernmental Panel on Climate Change (IPCC) Summary for Policymakers is for the water cycle to continue to intensify throughout the 21st century, though this does not mean that precipitation will increase in all regions. In subtropical land areas — places that are already relatively dry — precipitation is projected to decrease during the 21st century, increasing the probability of drought. The drying is projected to be strongest near the poleward margins of the subtropics (for example, the Mediterranean Basin, South Africa, southern Australia, and the Southwestern United States). Annual precipitation amounts are expected to increase in near-equatorial regions that tend to be wet in the present climate, and also at high latitudes. These large-scale patterns are present in nearly all of the climate model simulations conducted at several international research centers as part of the 4th Assessment of the IPCC. There is now ample evidence that increased hydrologic variability and change in climate has and will continue to have a profound impact on the water sector through the hydrologic cycle, water availability, water demand, and water allocation at the global, regional, basin, and local levels. Research published in 2012 in *Science* based on surface ocean salinity over the period 1950 to 2000 confirm this projection of an intensified global water cycle with salty areas becoming more saline and fresher areas becoming more fresh over the period:

Fundamental thermodynamics and climate models suggest that dry regions will become drier and wet regions will become wetter in response to warming. Efforts to detect this long-term response in sparse surface observations of rainfall and evaporation remain ambiguous. We show that ocean salinity patterns express an identifiable fingerprint of an intensifying water cycle. Our 50-year observed global surface salinity changes, combined with changes from global climate models, present robust evidence of an intensified global water cycle at a rate of 8 ± 5% per degree of surface warming. This rate is double the response projected by current-generation climate models and suggests that a substantial (16 to 24%) intensification of the global water cycle will occur in a future 2° to 3° warmer world.

An instrument carried by the SAC-D satellite launched in June, 2011 measures global sea surface salinity but data collection began only in June, 2011.

Glacial retreat is also an example of a changing water cycle, where the supply of water to glaciers from precipitation cannot keep up with the loss of water from melting and sublimation. Glacial retreat since 1850 has been extensive.

Human activities that alter the water cycle include:

- agriculture
- industry
- alteration of the chemical composition of the atmosphere
- construction of dams
- deforestation and afforestation
- removal of groundwater from wells
- water abstraction from rivers
- urbanization

Effects on Climate

The water cycle is powered from solar energy. 86% of the global evaporation occurs from the oceans, reducing their temperature by evaporative cooling. Without the cooling, the effect of evaporation on the greenhouse effect would lead to a much higher surface temperature of 67 °C (153 °F), and a warmer planet.

Aquifer drawdown or overdrafting and the pumping of fossil water increases the total amount of water in the hydrosphere, and has been postulated to be a contributor to sea-level rise.

Effects on Biogeochemical Cycling

While the water cycle is itself a biogeochemical cycle, flow of water over and beneath the Earth is a key component of the cycling of other biogeochemicals. Runoff is responsible for almost all of the transport of eroded sediment and phosphorus from land to waterbodies. The salinity of the oceans is derived from erosion and transport of dissolved salts from the land. Cultural eutrophication of lakes is primarily due to phosphorus, applied in excess to agricultural fields in fertilizers, and then transported overland and down rivers. Both runoff and groundwater flow play significant roles in transporting nitrogen from the land to waterbodies. The dead zone at the outlet of the Mississippi River is a consequence of nitrates from fertilizer being carried off agricultural fields and funnelled down the river system to the Gulf of Mexico. Runoff also plays a part in the carbon cycle, again through the transport of eroded rock and soil.

Slow Loss Over Geologic Time

The hydrodynamic wind within the upper portion of a planet's atmosphere allows light chemical elements such as Hydrogen to move up to the exobase, the lower limit of the exosphere, where the gases can then reach escape velocity, entering outer space without impacting other particles of gas. This type of gas loss from a planet into space is known as planetary wind. Planets with hot lower atmospheres could result in humid upper atmospheres that accelerate the loss of hydrogen.

History of Hydrologic Cycle Theory

Floating Land Mass

In ancient times, it was thought that the land mass floated on a body of water, and that most of the water in rivers has its origin under the earth. Examples of this belief can be found in the works of Homer (circa 800 BCE).

Source of Rain

In the ancient near east, Hebrew scholars observed that even though the rivers ran into the sea, the sea never became full (Ecclesiastes 1:7). Some scholars conclude that the water cycle was described completely during this time in this passage: "The wind goeth toward the south, and turneth about unto the north; it whirleth about continually, and the wind returneth again according to its circuits. All the rivers run into the sea; yet the sea is not full; unto the place from whence the rivers come, thither they return again" (Ecclesiastes 1:6-7, KJV). Scholars are not in agreement as to the date of Ecclesiastes, though most scholars point to a date during the time of Solomon, the son of David and Bathsheba, "three thousand years ago, there is some agreement that the time period is 962-922 BCE. Furthermore, it was also observed that when the clouds were full, they emptied rain on the earth (Ecclesiastes 11:3). In addition, during 793-740 BC a Hebrew prophet, Amos, stated that water comes from the sea and is poured out on the earth (Amos 5:8, 9:6).

Precipitation and Percolation

In the Adityahridayam (a devotional hymn to the Sun God) of Ramayana, a Hindu epic dated to the 4th century BC, it is mentioned in the 22nd verse that the Sun heats up water and sends it down as rain. By roughly 500 BCE, Greek scholars were speculating that much of the water in rivers can be attributed to rain. The origin of rain was also known by then. These scholars maintained the belief, however, that water rising up through the earth contributed a great deal to rivers. Examples of this thinking included Anaximander (570 BCE) (who also speculated about the evolution of land animals from fish) and Xenophanes of Colophon (530 BCE). Chinese scholars such as Chi Ni Tzu (320 BC) and Lu Shih Ch'un Ch'iu (239 BCE) had similar thoughts. The idea that the water cycle is a closed cycle can be found in the works of Anaxagoras of Clazomenae (460 BCE) and Diogenes of Apollonia (460 BCE). Both Plato (390 BCE) and Aristotle (350 BCE) speculated about percolation as part of the water cycle.

Precipitation Alone

In the Biblical Book of Job, dated between 7th and 2nd centuries BCE, there is a description of precipitation in the hydrologic cycle, "For he maketh small the drops of water: they pour down rain according to the vapour thereof; Which the clouds do drop and distil upon man abundantly" (Job 36:27-28, KJV). Also found in the book of Ecclesiastes "All the rivers flow into the sea, Yet the sea is not full. To the place where the rivers flow, There they flow again." (Ecclesiastes 1:7)

Up to the time of the Renaissance, it was thought that precipitation alone was insufficient to feed rivers, for a complete water cycle, and that underground water pushing upwards from the oceans were the main contributors to river water. Bartholomew of England held this view (1240 CE), as did Leonardo da Vinci (1500 CE) and Athanasius Kircher (1644 CE).

The first published thinker to assert that rainfall alone was sufficient for the maintenance of rivers was Bernard Palissy (1580 CE), who is often credited as the "discoverer" of the modern theory of the water cycle. Palissy's theories were not tested scientifically until 1674, in a study commonly attributed to Pierre Perrault. Even then, these beliefs were not accepted in mainstream science until the early nineteenth century.

Water Resources

Distribution of Earth's Water

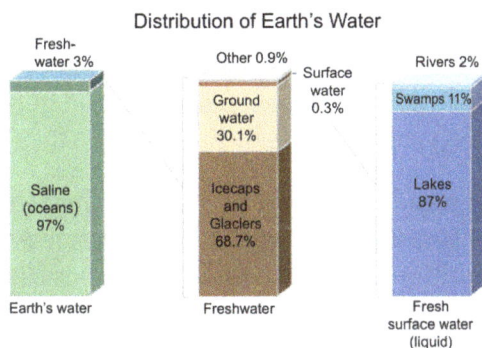

A graphical distribution of the locations of water on Earth. Only 3% of the Earth's water is fresh water. Most of it is in icecaps and glaciers (69%) and groundwater (30%), while all lakes, rivers and swamps combined only account for a small fraction (0.3%) of the Earth's total freshwater reserves.

Water resources are sources of water that are useful or potentially useful. Uses of water include agricultural, industrial, household, recreational and environmental activities. The majority of human uses require fresh water.

97% of the water on the Earth is salt water and only three percent is fresh water; slightly over two thirds of this is frozen in glaciers and polar ice caps. The remaining unfrozen freshwater is found mainly as groundwater, with only a small fraction present above ground or in the air.

Fresh water is a renewable resource, yet the world's supply of groundwater is steadily decreasing, with depletion occurring most prominently in Asia and North America, although it is still unclear how much natural renewal balances this usage, and whether ecosystems are threatened. The framework for allocating water resources to water users (where such a framework exists) is known as water rights.

Sources of Fresh Water

Surface Water

Lake Chungará and Parinacota volcano in northern Chile

Surface water is water in a river, lake or fresh water wetland. Surface water is naturally replenished by precipitation and naturally lost through discharge to the oceans, evaporation, evapotranspiration and groundwater recharge.

Although the only natural input to any surface water system is precipitation within its watershed, the total quantity of water in that system at any given time is also dependent on many other factors. These factors include storage capacity in lakes, wetlands and artificial reservoirs, the permeability of the soil beneath these storage bodies, the runoff characteristics of the land in the watershed, the timing of the precipitation and local evaporation rates. All of these factors also affect the proportions of water loss.

Human activities can have a large and sometimes devastating impact on these factors. Humans often increase storage capacity by constructing reservoirs and decrease it by draining wetlands. Humans often increase runoff quantities and velocities by paving areas and channelizing stream flow.

The total quantity of water available at any given time is an important consideration. Some human water users have an intermittent need for water. For example, many farms require large quantities of water in the spring, and no water at all in the winter. To supply such a farm with water, a surface water system may require a large storage capacity to collect water throughout the year and release it in a short period of time. Other users have a continuous need for water, such as a power plant that requires water for cooling. To supply such a power plant with water, a surface water system only needs enough storage capacity to fill in when average stream flow is below the power plant's need.

Nevertheless, over the long term the average rate of precipitation within a watershed is the upper bound for average consumption of natural surface water from that watershed.

Natural surface water can be augmented by importing surface water from another watershed through a canal or pipeline. It can also be artificially augmented from any of the other sources listed here, however in practice the quantities are negligible. Humans can also cause surface water to be "lost" (i.e. become unusable) through pollution.

Brazil is the country estimated to have the largest supply of fresh water in the world, followed by Russia and Canada.

Panorama of a natural wetland (Sinclair Wetlands, New Zealand)

Under River Flow

Throughout the course of a river, the total volume of water transported downstream will often be a combination of the visible free water flow together with a substantial contribution flowing through rocks and sediments that underlie the river and its floodplain called the hyporheic zone. For many rivers in large valleys, this unseen component of flow may greatly exceed the visible flow. The hyporheic zone often forms a dynamic interface between surface water and groundwater from aquifers, exchanging flow between rivers and aquifers that may be fully charged or depleted. This is especially significant in karst areas where pot-holes and underground rivers are common.

Groundwater

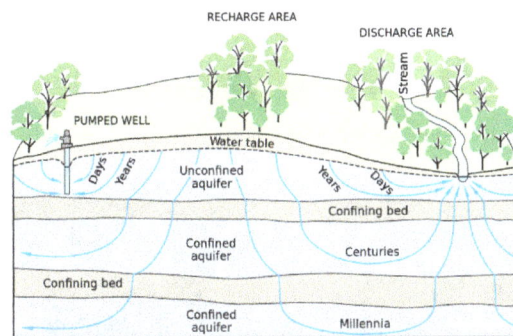

Relative groundwater travel times in the subsurface

Groundwater is fresh water located in the subsurface pore space of soil and rocks. It is also water that is flowing within aquifers below the water table. Sometimes it is useful to make a distinction between groundwater that is closely associated with surface water and deep groundwater in an aquifer (sometimes called "fossil water").

A shipot is a common water source in Central Ukrainian villages

Groundwater can be thought of in the same terms as surface water: inputs, outputs and storage. The critical difference is that due to its slow rate of turnover, groundwater storage is generally much larger (in volume) compared to inputs than it is for surface water. This difference makes it easy for humans to use groundwater unsustainably for a long time without severe consequences. Nevertheless, over the long term the average rate of seepage above a groundwater source is the upper bound for average consumption of water from that source.

The natural input to groundwater is seepage from surface water. The natural outputs from groundwater are springs and seepage to the oceans.

If the surface water source is also subject to substantial evaporation, a groundwater source may become saline. This situation can occur naturally under endorheic bodies of water, or artificially under irrigated farmland. In coastal areas, human use of a groundwater source may cause the direction of seepage to ocean to reverse which can also cause soil salinization. Humans can also cause groundwater to be "lost" (i.e. become unusable) through pollution. Humans can increase the input to a groundwater source by building reservoirs or detention ponds.

Frozen Water

Iceberg near Newfoundland

Several schemes have been proposed to make use of icebergs as a water source, however to date this has only been done for research purposes. Glacier runoff is considered to be surface water.

The Himalayas, which are often called "The Roof of the World", contain some of the most extensive and rough high altitude areas on Earth as well as the greatest area of glaciers and permafrost outside of the poles. Ten of Asia's largest rivers flow from there, and more than a billion people's livelihoods depend on them. To complicate matters, temperatures there are rising more rapidly than the global average. In Nepal, the temperature has risen by 0.6 degrees Celsius over the last decade, whereas globally, the Earth has warmed approximately 0.7 degrees Celsius over the last hundred years.

Desalination

Desalination is an artificial process by which saline water (generally sea water) is converted to fresh water. The most common desalination processes are distillation and reverse osmosis. Desalination is currently expensive compared to most alternative sources of water, and only a very small fraction of total human use is satisfied by desalination. It is only economically practical for high-valued uses (such as household and industrial uses) in arid areas. The most extensive use is in the Persian Gulf.

Water Uses

Agricultural

It is estimated that 70% of worldwide water is used for irrigation, with 15-35% of irrigation withdrawals being unsustainable. It takes around 2,000 - 3,000 litres of water to produce enough food to satisfy one person's daily dietary need. This is a considerable amount, when compared to that required for drinking, which is between two and five litres. To produce food for the now over 7 billion people who inhabit the planet today requires the water that would fill a canal ten metres deep, 100 metres wide and 2100 kilometres long.

Increasing Water Scarcity

Around fifty years ago, the common perception was that water was an infinite resource. At that time, there were fewer than half the current number of people on the planet. People were not as wealthy as today, consumed fewer calories and ate less meat, so less water was needed to produce their food. They required a third of the volume of water we presently take from rivers. Today, the competition for water resources is much more intense. This is because there are now seven billion people on the planet, their consumption of water-thirsty meat and vegetables is rising, and there is increasing competition for water from industry, urbanisation biofuel crops, and water reliant food items. In the future, even more water will be needed to produce food because the Earth's population is forecast to rise to 9 billion by 2050. An additional 2.5 or 3 billion people, choosing to eat fewer cereals and more meat and vegetables could add an additional five million kilometres to the virtual canal mentioned above.

An assessment of water management in agriculture sector was conducted in 2007 by the International Water Management Institute in Sri Lanka to see if the world had sufficient water to provide food for its growing population. It assessed the current availability of water for agriculture on a global scale and mapped out locations suffering from water scarcity. It found that a fifth of the world's people, more than 1.2 billion, live in areas of physical water scarcity, where there is not enough water to meet all demands. One third of the worlds population does not have access to clean drinking water, which is more than 2.3 billion people. A further 1.6 billion people live in areas experiencing economic water scarcity, where the lack of investment in water or insufficient human capacity make it impossible for authorities to satisfy the demand for water. The report found that it would be possible to produce the food required in future, but that continuation of today's food production and environmental trends would lead to crises in many parts of the world. To avoid a global water crisis, farmers will have to strive to increase productivity to meet growing demands for food, while industry and cities find ways to use water more efficiently.

In some areas of the world, irrigation is necessary to grow any crop at all, in other areas it permits more profitable crops to be grown or enhances crop yield. Various irrigation methods involve different trade-offs between crop yield, water consumption and capital cost of equipment and structures. Irrigation methods such as furrow and overhead sprinkler irrigation are usually less expensive but are also typically less efficient, because much of the water evaporates, runs off or drains below the root zone. Other irrigation methods considered to be more efficient include drip or trickle irrigation, surge irrigation, and some types of sprinkler systems where the sprinklers are operated near ground level. These types of systems, while more expensive, usually offer greater potential to minimize runoff, drainage and evaporation. Any system that is improperly managed can be wasteful, all methods have the potential for high efficiencies under suitable conditions, appropriate irrigation timing and management. Some issues that are often insufficiently considered are salinization of groundwater and contaminant accumulation leading to water quality declines.

As global populations grow, and as demand for food increases in a world with a fixed water supply, there are efforts under way to learn how to produce more food with less water, through improvements in irrigation methods and technologies, agricultural water management, crop types, and water monitoring. Aquaculture is a small but growing agricultural use of water. Freshwater commercial fisheries may also be considered as agricultural uses of water, but have generally been assigned a lower priority than irrigation.

Industrial

A power plant in Poland

It is estimated that 22% of worldwide water is used in industry. Major industrial users include hydroelectric dams, thermoelectric power plants, which use water for cooling, ore and oil refineries, which use water in chemical processes, and manufacturing plants, which use water as a solvent. Water withdrawal can be very high for certain industries, but consumption is generally much lower than that of agriculture.

Water is used in renewable power generation. Hydroelectric power derives energy from the force of water flowing downhill, driving a turbine connected to a generator. This hydroelectricity is a low-cost, non-polluting, renewable energy source. Significantly, hydroelectric power can also be used for load following unlike most renewable energy sources which are intermittent. Ultimately, the energy in a hydroelectric powerplant is supplied by the sun. Heat from the sun evaporates water, which condenses as rain in higher altitudes and flows downhill. Pumped-storage hydroelectric plants also exist, which use grid electricity to pump water uphill when demand is low, and use the stored water to produce electricity when demand is high.

Hydroelectric power plants generally require the creation of a large artificial lake. Evaporation from this lake is higher than evaporation from a river due to the larger surface area exposed to the elements, resulting in much higher water consumption. The process of driving water through the turbine and tunnels or pipes also briefly removes this water from the natural environment, creating water withdrawal. The impact of this withdrawal on wildlife varies greatly depending on the design of the powerplant.

Pressurized water is used in water blasting and water jet cutters. Also, very high pressure water guns are used for precise cutting. It works very well, is relatively safe, and is not harmful to the environment. It is also used in the cooling of machinery to prevent overheating, or prevent saw blades from overheating. This is generally a very small source of water consumption relative to other uses.

Water is also used in many large scale industrial processes, such as thermoelectric power production, oil refining, fertilizer production and other chemical plant use, and natural gas extraction from shale rock. Discharge of untreated water from industrial uses is pollution. Pollution includes discharged solutes (chemical pollution) and increased water temperature (thermal pollution). Industry requires pure water for many applications and utilizes a variety of purification techniques both in water supply and discharge. Most of this pure water is generated on site, either from natural freshwater or from municipal grey water. Industrial consumption of water is generally much lower than withdrawal, due to laws requiring industrial grey water to be treated and returned to the environment. Thermoelectric powerplants using cooling towers have high consumption, nearly equal to their withdrawal, as most of the withdrawn water is evaporated as part of the cooling process. The withdrawal, however, is lower than in once-through cooling systems.

Domestic

Drinking water

It is estimated that 8% of worldwide water use is for domestic purposes. These include drinking water, bathing, cooking, toilet flushing, cleaning, laundry and gardening. Basic domestic water requirements have been estimated by Peter Gleick at around 50 liters per person per day, excluding water for gardens. Drinking water is water that is of sufficiently high quality so that it can be consumed or used without risk of immediate or long term harm. Such water is commonly called potable water. In most developed countries, the water supplied to domestic, commerce and industry is all of drinking water standard even though only a very small proportion is actually consumed or used in food preparation.

Recreation

Whitewater rapids

Recreational water use is usually a very small but growing percentage of total water use. Recreational water use is mostly tied to reservoirs. If a reservoir is kept fuller than it would otherwise be for recreation, then the water retained could be categorized as recreational usage. Release of water from a few reservoirs is also timed to enhance whitewater boating, which also could be considered a recreational usage. Other examples are anglers, water skiers, nature enthusiasts and swimmers.

Recreational usage is usually non-consumptive. Golf courses are often targeted as using excessive amounts of water, especially in drier regions. It is, however, unclear whether recreational irrigation (which would include private gardens) has a noticeable effect on water resources. This is largely due to the unavailability of reliable data. Additionally, many golf courses utilize either primarily or exclusively treated effluent water, which has little impact on potable water availability.

Some governments, including the Californian Government, have labelled golf course usage as agricultural in order to deflect environmentalists' charges of wasting water. However, using the above figures as a basis, the actual statistical effect of this reassignment is close to zero. In Arizona, an organized lobby has been established in the form of the Golf Industry Association, a group focused on educating the public on how golf impacts the environment.

Recreational usage may reduce the availability of water for other users at specific times and places. For example, water retained in a reservoir to allow boating in the late summer is not available to farmers during the spring planting season. Water released for whitewater rafting may not be available for hydroelectric generation during the time of peak electrical demand.

Environmental

Explicit environment water use is also a very small but growing percentage of total water use. Environmental water may include water stored in impoundments and released for environmental purposes (held environmental water), but more often is water retained in waterways through regulatory limits of abstraction. Environmental water usage includes watering of natural or artificial wetlands, artificial lakes intended to create wildlife habitat, fish ladders, and water releases from reservoirs timed to help fish spawn, or to restore more natural flow regimes

Like recreational usage, environmental usage is non-consumptive but may reduce the availability of water for other users at specific times and places. For example, water release from a reservoir to help fish spawn may not be available to farms upstream, and water retained in a river to maintain waterway health would not be available to water abstractors downstream.

Water Stress

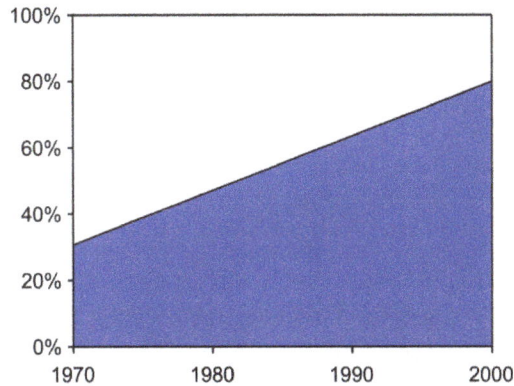

Estimate of the share of people in developing countries with access to drinking water 1970–2000

The concept of water stress is relatively simple: According to the World Business Council for Sustainable Development, it applies to situations where there is not enough water for all uses, whether agricultural, industrial or domestic. Defining thresholds for stress in terms of available water per capita is more complex, however, entailing assumptions about water use and its efficiency. Nevertheless, it has been proposed that when annual per capita renewable freshwater availability is less than 1,700 cubic meters, countries begin to experience periodic or regular water stress. Below 1,000 cubic meters, water scarcity begins to hamper economic development and human health and well-being.

Population Growth

In 2000, the world population was 6.2 billion. The UN estimates that by 2050 there will be an additional 3.5 billion people with most of the growth in developing countries that already suffer water stress. Thus, water demand will increase unless there are corresponding increases in water conservation and recycling of this vital resource. In building on the data presented here by the UN, the World Bank goes on to explain that access to water for producing food will be one of the main challenges in the decades to come. Access to water will need to be balanced with the importance of managing water itself in a sustainable way while taking into account the impact of climate change, and other environmental and social variables.

Expansion of Business Activity

Business activity ranging from industrialization to services such as tourism and entertainment continues to expand rapidly. This expansion requires increased water services including both supply and sanitation, which can lead to more pressure on water resources and natural ecosystem.

Rapid Urbanization

The trend towards urbanization is accelerating. Small private wells and septic tanks that work well in low-density communities are not feasible within high-density urban areas. Urbanization requires significant investment in water infrastructure in order to deliver water to individuals and to process the concentrations of wastewater – both from individuals and from business. These polluted and contaminated waters must be treated or they pose unacceptable public health risks.

In 60% of European cities with more than 100,000 people, groundwater is being used at a faster rate than it can be replenished. Even if some water remains available, it costs increasingly more to capture it.

Climate Change

Climate change could have significant impacts on water resources around the world because of the close connections between the climate and hydrological cycle. Rising temperatures will increase evaporation and lead to increases in precipitation, though there will be regional variations in rainfall. Both droughts and floods may become more frequent in different regions at different times, and dramatic changes in snowfall and snow melt are expected in mountainous areas. Higher temperatures will also affect water quality in ways that are not well understood. Possible impacts include increased eutrophication. Climate change could also mean an increase in demand for farm irrigation, garden sprinklers, and perhaps even swimming pools. There is now ample evidence that increased hydrologic variability and change in climate has and will continue have a profound impact on the water sector through the hydrologic cycle, water availability, water demand, and water allocation at the global, regional, basin, and local levels.

Depletion of Aquifers

Due to the expanding human population, competition for water is growing such that many of the world's major aquifers are becoming depleted. This is due both for direct human consumption as well as agricultural irrigation by groundwater. Millions of pumps of all sizes are currently extracting groundwater throughout the world. Irrigation in dry areas such as northern China, Nepal and India is supplied by groundwater, and is being extracted at an unsustainable rate. Cities that have experienced aquifer drops between 10 and 50 meters include Mexico City, Bangkok, Beijing, Madras and Shanghai.

Pollution and Water Protection

Water pollution is one of the main concerns of the world today. The governments of numerous countries have striven to find solutions to reduce this problem. Many pollutants threaten water supplies, but the most widespread, especially in developing countries, is the discharge of raw sewage into natural waters; this method of sewage disposal is the most common method in underdeveloped countries, but also is prevalent in quasi-developed countries such as China, India, Nepal and Iran. Sewage, sludge, garbage, and even toxic pollutants are all dumped into the water. Even if sewage is treated, problems still arise. Treated sewage forms sludge, which may be placed in landfills, spread out on land, incinerated or dumped at sea. In addition to sewage, nonpoint source pollution such as agricultural runoff is a significant source of pollution in some parts of the world, along with urban stormwater runoff and chemical wastes dumped by industries and governments.

Polluted water

Water and Conflicts

Competition for water has widely increased, and it has become more difficult to conciliate the necessities for water supply for human consumption, food production, ecosystems and other uses. Water administration is frequently involved in contradictory and complex problems. Approximately 10% of the worldwide annual runoff is used for human necessities. Several areas of the world are flooded, while others have such low precipitations that human life is almost impossible. As population and development increase, raising water demand, the possibility of problems inside a certain country or region increases, as it happens with others outside the region.

Over the past 25 years, politicians, academics and journalists have frequently predicted that disputes over water would be a source of future wars. Commonly cited quotes include: that of former Egyptian Foreign Minister and former Secretary-General of the United Nations Boutrous Ghali, who forecast, "The next war in the Middle East will be fought over water, not politics"; his successor at the UN, Kofi Annan, who in 2001 said, "Fierce competition for fresh water may well become a source of conflict and wars in the future," and the former Vice President of the World Bank, Ismail Serageldin, who said the wars of the next century will be over water unless significant changes in governance occurred. The water wars hypothesis had its roots in earlier research carried out on a small number of transboundary rivers such as the Indus, Jordan and Nile. These particular rivers became the focus because they had experienced water-related disputes. Specific events cited as evidence include Israel's bombing of Syria's attempts to divert the Jordan's headwaters, and military threats by Egypt against any country building dams in the upstream waters of the Nile. However, while some links made between conflict and water were valid, they did not necessarily represent the norm.

The only known example of an actual inter-state conflict over water took place between 2500 and 2350 BC between the Sumerian states of Lagash and Umma. Water stress has most often led to conflicts at local and regional levels. Tensions arise most often within national borders, in the downstream areas of distressed river basins. Areas such as the lower regions of China's Yellow

River or the Chao Phraya River in Thailand, for example, have already been experiencing water stress for several years. Water stress can also exacerbate conflicts and political tensions which are not directly caused by water. Gradual reductions over time in the quality and/or quantity of fresh water can add to the instability of a region by depleting the health of a population, obstructing economic development, and exacerbating larger conflicts.

Shared Water Resources Can Promote Collaboration

Water resources that span international boundaries are more likely to be a source of collaboration and cooperation than war. Scientists working at the International Water Management Institute have been investigating the evidence behind water war predictions. Their findings show that, while it is true there has been conflict related to water in a handful of international basins, in the rest of the world's approximately 300 shared basins the record has been largely positive. This is exemplified by the hundreds of treaties in place guiding equitable water use between nations sharing water resources. The institutions created by these agreements can, in fact, be one of the most important factors in ensuring cooperation rather than conflict.

The International Union for the Conservation of Nature (IUCN) published the book *Share: Managing water across boundaries*. One chapter covers the functions of trans-boundary institutions and how they can be designed to promote cooperation, overcome initial disputes and find ways of coping with the uncertainty created by climate change. It also covers how the effectiveness of such institutions can be monitored.

Water Shortages

In 2025, water shortages will be more prevalent among poorer countries where resources are limited and population growth is rapid, such as the Middle East, Africa, and parts of Asia. By 2025, large urban and peri-urban areas will require new infrastructure to provide safe water and adequate sanitation. This suggests growing conflicts with agricultural water users, who currently consume the majority of the water used by humans.

Generally speaking the more developed countries of North America, Europe and Russia will not see a serious threat to water supply by the year 2025, not only because of their relative wealth, but more importantly their populations will be better aligned with available water resources. North Africa, the Middle East, South Africa and northern China will face very severe water shortages due to physical scarcity and a condition of overpopulation relative to their carrying capacity with respect to water supply. Most of South America, Sub-Saharan Africa, Southern China and India will face water supply shortages by 2025; for these latter regions the causes of scarcity will be economic constraints to developing safe drinking water, as well as excessive population growth.

Economic Considerations

Water supply and sanitation require a huge amount of capital investment in infrastructure such as pipe networks, pumping stations and water treatment works. It is estimated that Organisation for Economic Co-operation and Development (OECD) nations need to invest at least USD 200 billion per year to replace aging water infrastructure to guarantee supply, reduce leakage rates and protect water quality.

International attention has focused upon the needs of the developing countries. To meet the Millennium Development Goals targets of halving the proportion of the population lacking access to safe drinking water and basic sanitation by 2015, current annual investment on the order of USD 10 to USD 15 billion would need to be roughly doubled. This does not include investments required for the maintenance of existing infrastructure.

Once infrastructure is in place, operating water supply and sanitation systems entails significant ongoing costs to cover personnel, energy, chemicals, maintenance and other expenses. The sources of money to meet these capital and operational costs are essentially either user fees, public funds or some combination of the two. An increasing dimension to consider is the flexibility of the water supply system.

Water Table

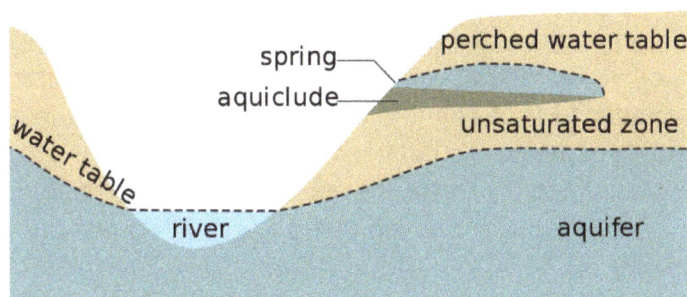

Cross section showing the water table varying with surface topography as well as a perched water table

The water table is the surface where the water pressure head is equal to the atmospheric pressure (where gauge pressure = 0). It may be conveniently visualized as the "surface" of the subsurface materials that are saturated with groundwater in a given vicinity. However, saturated conditions may extend above the water table as surface tension holds water in some pores below atmospheric pressure. Individual points on the water table are typically measured as the elevation that the water rises to in a well screened in the shallow groundwater.

The groundwater may be from infiltrating precipitation or from groundwater flowing into the aquifer. In areas with sufficient precipitation, water infiltrates through pore spaces in the soil, passing through the unsaturated zone. At increasing depths water fills in more of the pore spaces in the soils, until the zone of saturation is reached. In permeable or porous materials, such as sands and well fractured bedrock, the water table forms a relatively horizontal plane. Below the water table, in the phreatic zone, permeable units that yield groundwater are called aquifers. The ability of the aquifer to store groundwater is dependent on the primary and secondary porosity and permeability of the rock or soil. In less permeable soils, such as tight bedrock formations and historic lakebed deposits, the water table may be more difficult to define.

The water table should not be confused with the water level in a deeper well. If a deeper aquifer has a lower permeable unit that confines the upward flow, then the water level in a well screened in this aquifer may rise to a level that is greater or less than the elevation of the actual water table.

The elevation of the water in this deeper well is dependent upon the pressure in the deeper aquifer and is referred to as the potentiometric surface, not the water table.

Form

The water table may vary due to seasonal changes such as precipitation and evapotranspiration. In undeveloped regions with permeable soils that receive sufficient amounts of precipitation, the water table typically slopes toward rivers that act to drain the groundwater away and release the pressure in the aquifer. Springs, rivers, lakes and oases occur when the water table reaches the surface. Springs commonly form on hillsides, where the Earth's slanting surface may "intersect" with the water table. Groundwater entering rivers and lakes accounts for the base-flow water levels in water bodies.

Surface Topography

Within an aquifer, the water table is rarely horizontal, but reflects the surface relief due to the capillary effect in soils, sediments and other porous media. In the aquifer, groundwater flows from points of higher pressure to points of lower pressure, and the direction of groundwater flow typically has both a horizontal and a vertical component. The slope of the water table is known as the hydraulic gradient, which depends on the rate at which water is added to and removed from the aquifer and the permeability of the material. The water table does not always mimic the topography due to variations in the underlying geological structure (e.g., folded, faulted, fractured bedrock)

Perched Water Tables

A perched water table (or perched aquifer) is an aquifer that occurs above the regional water table, in the vadose zone. This occurs when there is an impermeable layer of rock or sediment (aquiclude) or relatively impermeable layer (aquitard) above the main water table/aquifer but below the surface of the land. If a perched aquifer's flow intersects the Earth's dry surface, at a valley wall for example, the water is discharged as a spring.

Fluctuations

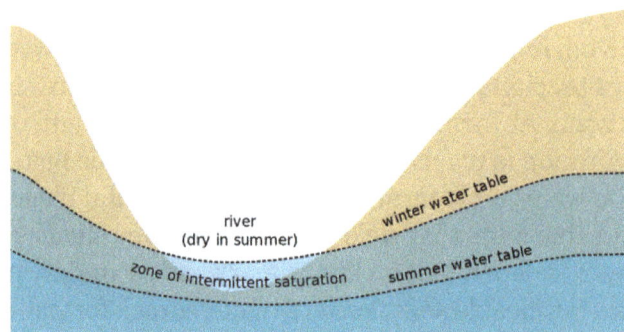

Seasonal fluctuations in the water table. During the dry season, river beds may dry up.

Tidal Fluctuations

On low-lying oceanic islands with porous soil, freshwater tends to collect in lenticular pools on top of the denser seawater intruding from the sides of the islands. Such an island's freshwater lens, and thus the water table, rises and falls with the tides.

Seasonal Fluctuations

In some regions, for example, Great Britain or California, winter precipitation is often higher than summer precipitation and so the groundwater storage is not fully recharged in summer. Consequently, the water table is lower during the summer. This disparity between the level of the winter and summer water table is known as the "zone of intermittent saturation", wherein the water table will fluctuate in response to climatic conditions.

Long-term Fluctuations

Fossil water is groundwater that has remained in an aquifer for several millennia and occurs mainly in deserts. It is non-renewable by present-day rainfall due to its depth below the surface, and any extraction causes a permanent change in the water table in such regions.

Effects on Climate

Aquifer drawdown or overdrafting and the pumping of fossil water may be a contributing factor to sea-level rise.

Drainage Basin

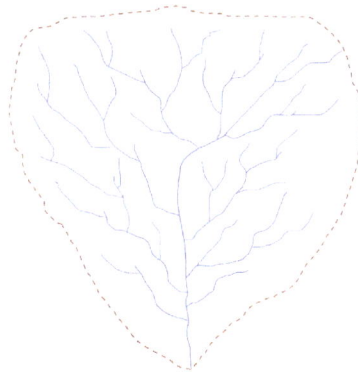

Example of a drainage basin. The dashed line is the main water divide of the hydrographic basin.

Other terms used to describe drainage basins are catchment, catchment area, drainage area, river basin and water basin. In North America, the term watershed is commonly used to mean a drainage basin, though in other English-speaking countries, it is used only in its original sense, to mean a drainage divide, the former meaning an area, the latter the high elevation perimeter of that area. Drainage basins drain into other drainage basins in a hierarchical pattern, with smaller sub-drainage basins combining into larger drainage basins.

In closed ("endorheic") drainage basins the water converges to a single point inside the basin, known as a sink, which may be a permanent lake, a dry lake, or a point where surface water is lost underground. The drainage basin includes all the streams and rivers that convey the water towards the sink, as well as the land surfaces from which water drains into those channels.

Latoriţa River, tributary of the Lotru River

(Drainage basin)

A drainage basin or catchment basin is an extent or an area of land where all surface water from rain, melting snow, or ice converges to a single point at a lower elevation, usually the exit of the basin, where the waters join another body of water, such as a river, lake, reservoir, estuary, wetland, sea, or ocean. Thus if a tributary stream joins a brook that in turn joins a small river which is a tributary of a larger river, there is a series of successively larger (and lower elevation) drainage basins. For instance, the Missouri and Ohio rivers are within their own drainage basins and also within the drainage basin of the Mississippi River.

The drainage basin acts as a funnel by collecting all the water within the area covered by the basin and channelling it to a single point. Each drainage basin is separated topographically from adjacent basins by a perimeter, the drainage divide, making up a succession of higher geographical features (such as a ridge, hill or mountains) forming a barrier.

Drainage basins are similar but not identical to hydrologic units, which are drainage areas delineated so as to nest into a multi-level hierarchical drainage system. Hydrologic units are defined to allow multiple inlets, outlets, or sinks. In a strict sense, all drainage basins are hydrologic units but not all hydrologic units are drainage basins.

Major Drainage Basins of the World

Map

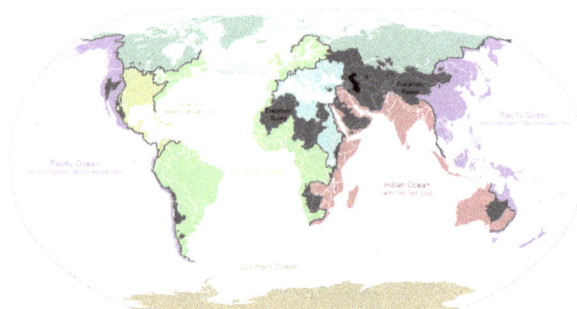

Drainage basins of the principal oceans and seas of the world. Grey areas are endorheic basins that do not drain to the ocean.

Ocean Basins

The following is a list of the major ocean basins:

- About 48.7% of the world's land drains to the Atlantic Ocean. In North America, surface water drains to the Atlantic via the Saint Lawrence River and Great Lakes basins, the Eastern Seaboard of the United States, the Canadian Maritimes, and most of Newfoundland and Labrador. Nearly all of South America east of the Andes also drains to the Atlantic, as does most of Western and Central Europe and the greatest portion of western Sub-Saharan Africa, as well as Western Sahara and part of Morocco. The two major mediterranean seas of the world also flow to the Atlantic:

    o The Caribbean Sea and Gulf of Mexico basin includes most of the U.S. interior between the Appalachian and Rocky Mountains, a small part of the Canadian provinces of Alberta and Saskatchewan, eastern Central America, the islands of the Caribbean and the Gulf, and a small part of northern South America.

    o The Mediterranean Sea basin includes much of North Africa, east-central Africa (through the Nile River), Southern, Central, and Eastern Europe, Turkey, and the coastal areas of Israel, Lebanon, and Syria.

- The Arctic Ocean drains most of Western and Northern Canada east of the Continental Divide, northern Alaska and parts of North Dakota, South Dakota, Minnesota, and Montana in the United States, the north shore of the Scandinavian peninsula in Europe, and much of central and northern Russia, and parts of Kazakhstan and Mongolia in Asia, which totals to about 17% of the world's land.

- Just over 13% of the land in the world drains to the Pacific Ocean. Its basin includes much of China, eastern and southeastern Russia, Japan, the Korean Peninsula, most of Indochina, Indonesia and Malaysia, the Philippines, all of the Pacific Islands, the northeast coast of Australia, and Canada and the United States west of the Continental Divide (including most of Alaska), as well as western Central America and South America west of the Andes.

- The Indian Ocean's drainage basin also comprises about 13% of Earth's land. It drains the eastern coast of Africa, the coasts of the Red Sea and the Persian Gulf, the Indian subcontinent, Burma, and most of Australia.

- The Southern Ocean drains Antarctica. Antarctica comprises approximately eight percent of the Earth's land.

Largest River Basins

The five largest river basins (by area), from largest to smallest, are the basins of the Amazon, the Río de la Plata, the Congo, the Nile, and the Mississippi. The three rivers that drain the most water, from most to least, are the Amazon, Ganga, and Congo rivers.

Endorheic Drainage Basins

Endorheic drainage basins are inland basins that do not drain to an ocean. Around 18% of all land drains to endorheic lakes or seas or sinks. The largest of these consists of much of the interior of Asia, which drains into the Caspian Sea, the Aral Sea, and numerous smaller lakes. Other endorheic regions include the Great Basin in the United States, much of the Sahara Desert, the drainage basin of the Okavango River (Kalahari Basin), highlands near the African Great Lakes, the interiors of Australia and the Arabian Peninsula, and parts in Mexico and the Andes. Some of these, such as the Great Basin, are not single drainage basins but collections of separate, adjacent closed basins.

Endorheic basin in Central Asia

In endorheic bodies of standing water where evaporation is the primary means of water loss, the water is typically more saline than the oceans. An extreme example of this is the Dead Sea.

Importance of Drainage Basins

Geopolitical Boundaries

Drainage basins have been historically important for determining territorial boundaries, particularly in regions where trade by water has been important. For example, the English crown gave the Hudson's Bay Company a monopoly on the fur trade in the entire Hudson Bay basin, an area called Rupert's Land. Bioregional political organization today includes agreements of states (e.g., international treaties and, within the U.S.A., interstate compacts) or other political entities in a particular drainage basin to manage the body or bodies of water into which it drains. Examples of such interstate compacts are the Great Lakes Commission and the Tahoe Regional Planning Agency.

Hydrology

Drainage basin of the Ohio River, part of the Mississippi River drainage basin

In hydrology, the drainage basin is a logical unit of focus for studying the movement of water within the hydrological cycle, because the majority of water that discharges from the basin outlet originated as precipitation falling on the basin. A portion of the water that enters the groundwater system beneath the drainage basin may flow towards the outlet of another drainage basin because groundwater flow directions do not always match those of their overlying drainage network. Measurement of the discharge of water from a basin may be made by a stream gauge located at the basin's outlet.

Rain gauge data is used to measure total precipitation over a drainage basin, and there are different ways to interpret that data. If the gauges are many and evenly distributed over an area of uniform precipitation, using the arithmetic mean method will give good results. In the Thiessen polygon method, the drainage basin is divided into polygons with the rain gauge in the middle of each polygon assumed to be representative for the rainfall on the area of land included in its polygon. These polygons are made by drawing lines between gauges, then making perpendicular bisectors of those lines form the polygons. The isohyetal method involves contours of equal precipitation are drawn over the gauges on a map. Calculating the area between these curves and adding up the volume of water is time consuming.

Isochrone maps can be used to show the time taken for runoff water within a drainage basin to reach a lake, reservoir or outlet, assuming constant and uniform effective rainfall.

Geomorphology

Drainage basins are the principal hydrologic unit considered in fluvial geomorphology. A drainage basin is the source for water and sediment that moves through the river system and reshapes the channel.

Ecology

The Mississippi River drains the largest area of any U.S. river, much of it agricultural regions. Agricultural runoff and other water pollution that flows to the outlet is the cause of the hypoxic, or dead zone in the Gulf of Mexico.

Drainage basins are important in ecology. As water flows over the ground and along rivers it can pick up nutrients, sediment, and pollutants. With the water, they are transported towards the outlet of the basin, and can affect the ecological processes along the way as well as in the receiving water source.

Modern use of artificial fertilizers, containing nitrogen, phosphorus, and potassium, has affected the mouths of drainage basins. The minerals are carried by the drainage basin to the mouth, and may accumulate there, disturbing the natural mineral balance. This can cause eutrophication where plant growth is accelerated by the additional material.

Resource Management

Because drainage basins are coherent entities in a hydrological sense, it has become common to manage water resources on the basis of individual basins. In the U.S. state of Minnesota, governmental entities that perform this function are called "watershed districts". In New Zealand, they are called catchment boards. Comparable community groups based in Ontario, Canada, are called conservation authorities. In North America this function is referred to as "watershed management". In Brazil, the National Policy of Water Resources, regulated by Act n° 9.433 of 1997, establishes the drainage basin as territorial division of Brazilian water management.

Catchment Factors

The catchment is the most significant factor determining the amount or likelihood of flooding.

Catchment factors are: topography, shape, size, soil type, and land use (paved or roofed areas). Catchment topography and shape determine the time taken for rain to reach the river, while catchment size, soil type, and development determine the amount of water to reach the river.

Topography

Generally, topography plays a big part in how fast runoff will reach a river. Rain that falls in steep mountainous areas will reach the primary river in the drainage basin faster than flat or lightly sloping areas (e.g., > 1% gradient).

Shape

Shape will contribute to the speed with which the runoff reaches a river. A long thin catchment will take longer to drain than a circular catchment.

Size

Size will help determine the amount of water reaching the river, as the larger the catchment the greater the potential for flooding. It also determined on the basis of length and width of the drainage basin.

Soil Type

Soil type will help determine how much water reaches the river. Certain soil types such as sandy soils are very free-draining, and rainfall on sandy soil is likely to be absorbed by the ground. However, soils containing clay can be almost impermeable and therefore rainfall on clay soils will run off and contribute to flood volumes. After prolonged rainfall even free-draining soils can become saturated, meaning that any further rainfall will reach the river rather than being absorbed by the ground. If the surface is impermeable the precipitation will create surface run-off which will lead to higher risk of flooding; if the ground is permeable, the precipitation will infiltrate the soil.

Land Use

Land use can contribute to the volume of water reaching the river, in a similar way to clay soils. For example, rainfall on roofs, pavements, and roads will be collected by rivers with almost no absorption into the groundwater.

Groundwater Flow

In hydrogeology, groundwater flow is defined as the "...part of streamflow that has infiltrated the ground, has entered the phreatic zone, and has been discharged into a stream channel, via springs or seepage water". It is governed by the groundwater flow equation. Groundwater is water that is found underground in cracks and spaces in the soil, sand and rocks. An area where water fills these spaces is called a phreatic zone or saturated zone. Groundwater is stored in and moves slowly through the layers of soil, sand and rocks called aquifers. The rate of groundwater flow depends on the permeability (the size of the spaces in the soil or rocks and how well the spaces are connected) and the hydraulic head (water pressure).

Aquifer

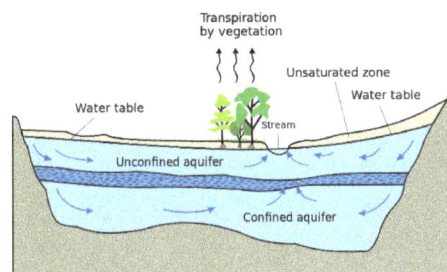

Typical aquifer cross-section

An aquifer is an underground layer of water-bearing permeable rock, rock fractures or unconsolidated materials (gravel, sand, or silt) from which groundwater can be extracted using a water well. The study of water flow in aquifers and the characterization of aquifers is called hydrogeology. Related terms include aquitard, which is a bed of low permeability along an aquifer, and aquiclude (or *aquifuge*), which is a solid, impermeable area underlying or overlying an aquifer. If the impermeable area overlies the aquifer, pressure could cause it to become a confined aquifer.

Depth

Aquifers may occur at various depths. Those closer to the surface are not only more likely to be used for water supply and irrigation, but are also more likely to be topped up by the local rainfall. Many desert areas have limestone hills or mountains within them or close to them that can be exploited as groundwater resources. Part of the Atlas Mountains in North Africa, the Lebanon and Anti-Lebanon ranges between Syria and Lebanon, the Jebel Akhdar (Oman) in Oman, parts of the Sierra Nevada and neighboring ranges in the United States' Southwest, have shallow aquifers that are exploited for their water. Overexploitation can lead to the exceeding of the practical sustained yield; i.e., more water is taken out than can be replenished. Along the coastlines of certain countries, such as Libya and Israel, increased water usage associated with population growth has caused a lowering of the water table and the subsequent contamination of the groundwater with saltwater from the sea.

The beach provides a model to help visualize an aquifer. If a hole is dug into the sand, very wet or saturated sand will be located at a shallow depth. This hole is a crude well, the wet sand represents an aquifer, and the level to which the water rises in this hole represents the water table.

In 2013 large freshwater aquifers were discovered under continental shelves off Australia, China, North America and South Africa. They contain an estimated half a million cubic kilometers of "low salinity" water that could be economically processed into potable water. The reserves formed when ocean levels were lower and rainwater made its way into the ground in land areas that were not submerged until the ice age ended 20,000 years ago. The volume is estimated to be 100x the amount of water extracted from other aquifers since 1900.

Classification

The above diagram indicates typical flow directions in a cross-sectional view of a simple confined or unconfined aquifer system. The system shows two aquifers with one aquitard (a confining or impermeable layer) between them, surrounded by the bedrock *aquiclude*, which is in contact with a gaining stream (typical in humid regions). The water table and unsaturated zone are also illustrated. An *aquitard* is a zone within the earth that restricts the flow of groundwater from one aquifer to another. An aquitard can sometimes, if completely impermeable, be called an *aquiclude* or *aquifuge*. Aquitards are composed of layers of either clay or non-porous rock with low hydraulic conductivity.

Saturated Versus Unsaturated

Groundwater can be found at nearly every point in the Earth's shallow subsurface to some degree, although aquifers do not necessarily contain fresh water. The Earth's crust can be divided into two regions: the *saturated zone* or *phreatic zone* (e.g., aquifers, aquitards, etc.), where all available spaces are filled with water, and the *unsaturated zone* (also called the vadose zone), where there are still pockets of air that contain some water, but can be filled with more water.

Saturated means the pressure head of the water is greater than atmospheric pressure (it has a gauge pressure > 0). The definition of the water table is the surface where the pressure head is equal to atmospheric pressure (where gauge pressure = 0).

Unsaturated conditions occur above the water table where the pressure head is negative (absolute pressure can never be negative, but gauge pressure can) and the water that incompletely fills the pores of the aquifer material is under suction. The water content in the unsaturated zone is held in place by surface adhesive forces and it rises above the water table (the zero-gauge-pressure isobar) by capillary action to saturate a small zone above the phreatic surface (the capillary fringe) at less than atmospheric pressure. This is termed tension saturation and is not the same as saturation on a water-content basis. Water content in a capillary fringe decreases with increasing distance from the phreatic surface. The capillary head depends on soil pore size. In sandy soils with larger pores, the head will be less than in clay soils with very small pores. The normal capillary rise in a clayey soil is less than 1.80 m (six feet) but can range between 0.3 and 10 m (one and 30 ft).

The capillary rise of water in a small-diameter tube involves the same physical process. The water table is the level to which water will rise in a large-diameter pipe (e.g., a well) that goes down into the aquifer and is open to the atmosphere.

Aquifers Versus Aquitards

Aquifers are typically saturated regions of the subsurface that produce an economically feasible quantity of water to a well or spring (e.g., sand and gravel or fractured bedrock often make good aquifer materials).

An aquitard is a zone within the earth that restricts the flow of groundwater from one aquifer to another. A completely impermeable aquitard is called an aquiclude or aquifuge. Aquitards comprise layers of either clay or non-porous rock with low hydraulic conductivity.

In mountainous areas (or near rivers in mountainous areas), the main aquifers are typically unconsolidated alluvium, composed of mostly horizontal layers of materials deposited by water processes (rivers and streams), which in cross-section (looking at a two-dimensional slice of the aquifer) appear to be layers of alternating coarse and fine materials. Coarse materials, because of the high energy needed to move them, tend to be found nearer the source (mountain fronts or rivers), whereas the fine-grained material will make it farther from the source (to the flatter parts of the basin or overbank areas - sometimes called the pressure area). Since there are less fine-grained deposits near the source, this is a place where aquifers are often unconfined (sometimes called the forebay area), or in hydraulic communication with the land surface.

Confined Versus Unconfined

There are two end members in the spectrum of types of aquifers; *confined* and *unconfined* (with semi-confined being in between). Unconfined aquifers are sometimes also called *water table* or *phreatic* aquifers, because their upper boundary is the water table or phreatic surface. Typically (but not always) the shallowest aquifer at a given location is unconfined, meaning it does not have a confining layer (an aquitard or aquiclude) between it and the surface. The term "perched" refers to ground water accumulating above a low-permeability unit or strata, such as a clay layer. This term is generally used to refer to a small local area of ground water that occurs at an elevation higher than a regionally extensive aquifer. The difference between perched and unconfined aquifers is their size (perched is smaller). Confined aquifers are aquifers that are overlain by a confining layer, often made up of clay. The confining layer might offer some protection from surface contamination.

If the distinction between confined and unconfined is not clear geologically (i.e., if it is not known if a clear confining layer exists, or if the geology is more complex, e.g., a fractured bedrock aquifer), the value of storativity returned from an aquifer test can be used to determine it (although aquifer tests in unconfined aquifers should be interpreted differently than confined ones). Confined aquifers have very low storativity values (much less than 0.01, and as little as 10^{-5}), which means that the aquifer is storing water using the mechanisms of aquifer matrix expansion and the compressibility of water, which typically are both quite small quantities. Unconfined aquifers have storativities (typically then called specific yield) greater than 0.01 (1% of bulk volume); they release water from storage by the mechanism of actually draining the pores of the aquifer, releasing relatively large amounts of water (up to the drainable porosity of the aquifer material, or the minimum volumetric water content).

Isotropic Versus Anisotropic

In isotropic aquifers or aquifer layers the hydraulic conductivity (K) is equal for flow in all directions, while in anisotropic conditions it differs, notably in horizontal (Kh) and vertical (Kv) sense.

Semi-confined aquifers with one or more aquitards work as an anisotropic system, even when the separate layers are isotropic, because the compound Kh and Kv values are different.

When calculating flow to drains or flow to wells in an aquifer, the anisotropy is to be taken into account lest the resulting design of the drainage system may be faulty.

Groundwater in Rock Formations

Groundwater may exist in *underground rivers* (e.g., caves where water flows freely underground). This may occur in eroded limestone areas known as karst topography, which make up only a small percentage of Earth's area. More usual is that the pore spaces of rocks in the subsurface are simply saturated with water — like a kitchen sponge — which can be pumped out for agricultural, industrial, or municipal uses.

If a rock unit of low porosity is highly fractured, it can also make a good aquifer (via fissure flow), provided the rock has a hydraulic conductivity sufficient to facilitate movement of water. Porosity is important, but, *alone*, it does not determine a rock's ability to act as an aquifer. Areas of the Deccan Traps (a basaltic lava) in west central India are good examples of rock formations with high porosity but low permeability, which makes them poor aquifers. Similarly, the micro-porous (Upper Cretaceous) Chalk of south east England, although having a reasonably high porosity, has a low grain-to-grain permeability, with its good water-yielding characteristics mostly due to micro-fracturing and fissuring.

Human Dependence on Groundwater

Most land areas on Earth have some form of aquifer underlying them, sometimes at significant depths. In some cases, these aquifers are rapidly being depleted by the human population.

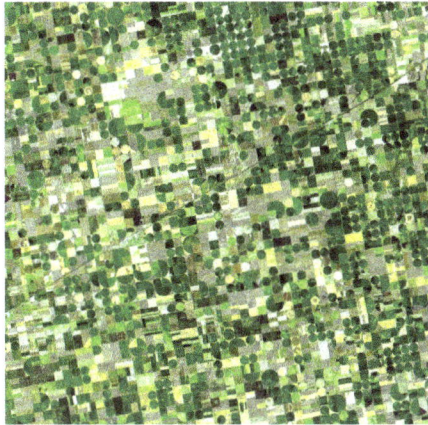

Center-pivot irrigated fields in Kansas covering hundreds of square miles watered by the Ogallala Aquifer

Fresh-water aquifers, especially those with limited recharge by snow or rain, also known as meteoric water, can be over-exploited and depending on the local hydrogeology, may draw in non-potable water or saltwater intrusion from hydraulically connected aquifers or surface water bodies. This can be a serious problem, especially in coastal areas and other areas where aquifer pumping is excessive. In some areas, the ground water can become contaminated by arsenic and other mineral poisons.

Aquifers are critically important in human habitation and agriculture. Deep aquifers in arid areas have long been water sources for irrigation. Many villages and even large cities draw their water supply from wells in aquifers.

Municipal, irrigation, and industrial water supplies are provided through large wells. Multiple wells for one water supply source are termed "wellfields", which may withdraw water from confined or unconfined aquifers. Using ground water from deep, confined aquifers provides more protection from surface water contamination. Some wells, termed "collector wells," are specifically designed to induce infiltration of surface (usually river) water.

Aquifers that provide sustainable fresh groundwater to urban areas and for agricultural irrigation are typically close to the ground surface (within a couple of hundred metres) and have some recharge by fresh water. This recharge is typically from rivers or meteoric water (precipitation) that percolates into the aquifer through overlying unsaturated materials.

Occasionally, sedimentary or "fossil" aquifers are used to provide irrigation and drinking water to urban areas. In Libya, for example, Muammar Gaddafi's Great Manmade River project has pumped large amounts of groundwater from aquifers beneath the Sahara to populous areas near the coast. Though this has saved Libya money over the alternative, desalination, the aquifers are likely to run dry in 60 to 100 years. Aquifer depletion has been cited as one of the causes of the food price rises of 2011.

Subsidence

In unconsolidated aquifers, groundwater is produced from pore spaces between particles of gravel, sand, and silt. If the aquifer is confined by low-permeability layers, the reduced water pressure in the sand and gravel causes slow drainage of water from the adjoining confining layers. If these confining layers are composed of compressible silt or clay, the loss of water to the aquifer reduces the water pressure in the confining layer, causing it to compress from the weight of overlying geologic materials. In severe cases, this compression can be observed on the ground surface as subsidence. Unfortunately, much of the subsidence from groundwater extraction is permanent (elastic rebound is small). Thus, the subsidence is not only permanent, but the compressed aquifer has a permanently reduced capacity to hold water.

Saltwater Intrusion

Aquifers near the coast have a lens of freshwater near the surface and denser seawater under freshwater. Seawater penetrates the aquifer diffusing in from the ocean and is denser than freshwater. For porous (i.e., sandy) aquifers near the coast, the thickness of freshwater atop saltwater is about 40 feet (12 m) for every 1 ft (0.30 m) of freshwater head above sea level. This relationship is called the Ghyben-Herzberg equation. If too much ground water is pumped near the coast, salt-water may intrude into freshwater aquifers causing contamination of potable freshwater supplies. Many coastal aquifers, such as the Biscayne Aquifer near Miami and the New Jersey Coastal Plain aquifer, have problems with saltwater intrusion as a result of overpumping and sea level rise.

Salination

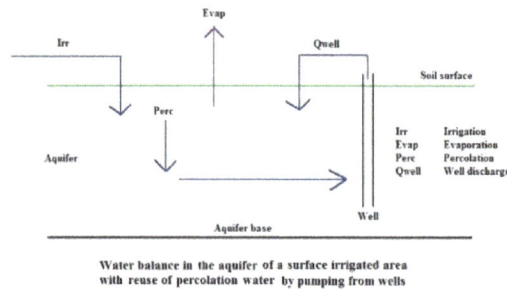

Water balance in the aquifer of a surface irrigated area
with reuse of percolation water by pumping from wells

Diagram of a water balance of the aquifer

Aquifers in surface irrigated areas in semi-arid zones with reuse of the unavoidable irrigation water losses percolating down into the underground by supplemental irrigation from wells run the risk of salination.

Surface irrigation water normally contains salts in the order of 0.5 g/l or more and the annual irrigation requirement is in the order of 10000 m³/ha or more so the annual import of salt is in the order of 5000 kg/ha or more.

Under the influence of continuous evaporation, the salt concentration of the aquifer water may increase continually and eventually cause an environmental problem.

For salinity control in such a case, annually an amount of drainage water is to be discharged from the aquifer by means of a subsurface drainage system and disposed of through a safe outlet. The drainage system may be *horizontal* (i.e. using pipes, tile drains or ditches) or *vertical* (drainage by wells). To estimate the drainage requirement, the use of a groundwater model with an agro-hydro-salinity component may be instrumental, e.g. SahysMod.

Examples

The Great Artesian Basin situated in Australia is arguably the largest groundwater aquifer in the world (over 1.7 million km²). It plays a large part in water supplies for Queensland and remote parts of South Australia.

The Guarani Aquifer, located beneath the surface of Argentina, Brazil, Paraguay, and Uruguay, is one of the world's largest aquifer systems and is an important source of fresh water. Named after the Guarani people, it covers 1,200,000 km², with a volume of about 40,000 km³, a thickness of between 50 m and 800 m and a maximum depth of about 1,800 m.

Aquifer depletion is a problem in some areas, and is especially critical in northern Africa; see the Great Manmade River project of Libya for an example. However, new methods of ground-water management such as artificial recharge and injection of surface waters during seasonal wet periods has extended the life of many freshwater aquifers, especially in the United States.

The Ogallala Aquifer of the central United States is one of the world's great aquifers, but in places it is being rapidly depleted by growing municipal use, and continuing agricultural use. This huge aquifer, which underlies portions of eight states, contains primarily fossil water from the time of the last glaciation. Annual recharge, in the more arid parts of the aquifer, is estimated to total only about 10 percent of annual withdrawals. According to a 2013 report by research hydrologist Leonard F. Konikow at the United States Geological Survey (USGC), the depletion between 2001–2008, inclusive, is about 32 percent of the cumulative depletion during the entire 20th century (Konikow 2013:22)." In the United States, the biggest users of water from aquifers include agricultural irrigation and oil and coal extraction. "Cumulative total groundwater depletion in the United States accelerated in the late 1940s and continued at an almost steady linear rate through the end of the century. In addition to widely recognized environmental consequences, groundwater depletion also adversely impacts the long-term sustainability of groundwater supplies to help meet the Nation's water needs."

An example of a significant and sustainable carbonate aquifer is the Edwards Aquifer in central Texas. This carbonate aquifer has historically been providing high quality water for nearly 2 million people, and even today, is full because of tremendous recharge from a number of area streams, rivers and lakes. The primary risk to this resource is human development over the recharge areas.

Discontinuous sand bodies at the base of the McMurray Formation in the Athabasca Oil Sands region of northeastern Alberta, Canada, are commonly referred to as the Basal Water Sand (BWS) aquifers. Saturated with water, they are confined beneath impermeable bitumen-saturated sands that are exploited to recover bitumen for synthetic crude oil production. Where they are deep-lying and recharge occurs from underlying Devonian formations they are saline, and where they are shallow and recharged by meteoric water they are non-saline. The BWS typically pose problems for the recovery of bitumen, whether by open-pit mining or by *in situ* methods such as steam-assisted gravity drainage (SAGD), and in some areas they are targets for waste-water injection.

Hydraulic Conductivity

Hydraulic conductivity, symbolically represented as K, is a property of vascular plants, soils and rocks, that describes the ease with which a fluid (usually water) can move through pore spaces or fractures. It depends on the intrinsic permeability of the material, the degree of saturation, and on the density and viscosity of the fluid. Saturated hydraulic conductivity, K_{sat}, describes water movement through saturated media.

By definition, hydraulic conductivity is the ratio of velocity to hydraulic gradient indicating permeability of porous media.

Methods of Determination

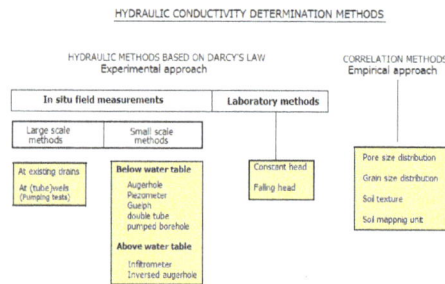

Overview of determination methods

There are two broad categories of determining hydraulic conductivity:

- *Empirical* approach by which the hydraulic conductivity is correlated to soil properties like pore size and particle size (grain size) distributions, and soil texture

- *Experimental* approach by which the hydraulic conductivity is determined from hydraulic experiments using Darcy's law

The experimental approach is broadly classified into:

- Laboratory tests using soil samples subjected to hydraulic experiments

- Field tests (on site, in situ) that are differentiated into:

    o small scale field tests, using observations of the water level in cavities in the soil

    o large scale field tests, like pump tests in wells or by observing the functioning of existing horizontal drainage systems.

The small scale field tests are further subdivided into:

- infiltration tests in cavities above the water table

- slug tests in cavities below the water table

Estimation By Empirical Approach

Estimation from Grain Size

Allen Hazen derived an empirical formula for approximating hydraulic conductivity from grain size analyses:

$$K = C(D_{10})^2$$

where

C Hazen's empirical coefficient, which takes a value between 0.0 and 1.5 (depending on literatures), with an average value of 1.0. A.F. Salarashayeri & M. Siosemarde give C as usually taken between 1.0 and 1.5, with D in mm and K in cm/s.

D_{10} is the diameter of the 10 percentile grain size of the material

Pedotransfer Function

A pedotransfer function (PTF) is a specialized empirical estimation method, used primarily in the soil sciences, however has increasing use in hydrogeology. There are many different PTF methods, however, they all attempt to determine soil properties, such as hydraulic conductivity, given several measured soil properties, such as soil particle size, and bulk density.

Determination by Experimental Approach

There are relatively simple and inexpensive laboratory tests that may be run to determine the hydraulic conductivity of a soil: constant-head method and falling-head method.

Laboratory Methods

Constant-head Method

The constant-head method is typically used on granular soil. This procedure allows water to move through the soil under a steady state head condition while the quantity (volume) of water flowing through the soil specimen is measured over a period of time. By knowing the quantity Q of water measured, length L of specimen, cross-sectional area A of the specimen, time t required for the quantity of water Q to be discharged, and head h, the hydraulic conductivity can be calculated:

$$\frac{Q}{t} = Av$$

where v is the flow velocity. Using Darcy's Law:

$$v = Ki$$

and expressing the hydraulic gradient i as:

$$i = \frac{h}{L}$$

where h is the difference of hydraulic head over distance L, yields:

$$Q = \frac{AKh}{L}$$

Solving for K gives:

$$K = \frac{QL}{Ah}$$

Falling-head Method

In the falling-head method, the soil sample is first saturated under a specific head condition. The water is then allowed to flow through the soil without adding any water, so the pressure head declines as water passes through the specimen. The advantage to the falling-head method is that it can be used for both fine-grained and coarse-grained soils. Calculating the hydraulic conductivity is more complicated because of the changing pressure head, and requires solving a differential equation; the resulting equation is:

$$K = \frac{2.3aL}{At} \log\left(\frac{h_1}{h_2}\right)$$

In-situ (Field) Methods

Augerhole Method

There are also in-situ methods for measuring the hydraulic conductivity in the field. When the water table is shallow, the augerhole method, a slug test, can be used for determining the hydraulic conductivity below the water table. The method was developed by Hooghoudt (1934) in The Netherlands and introduced in the US by Van Bavel en Kirkham (1948). The method uses the following steps:

1. an augerhole is perforated into the soil to below the water table

2. water is bailed out from the augerhole

3. the rate of rise of the water level in the hole is recorded

4. the K-value is calculated from the data as:

    K = F (Ho-Ht) / t

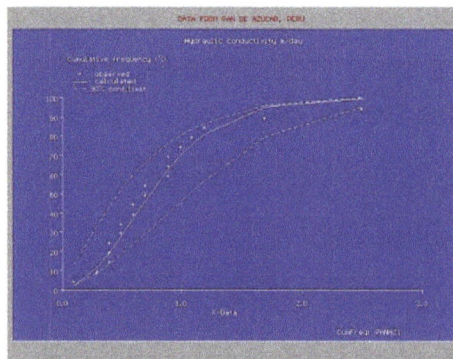

Cumulative frequency distribution (lognormal) of hydraulic conductivity (X-data)

where: K = horizontal saturated hydraulic conductivity (m/day), H = depth of the waterlevel in the hole relative to the water table in the soil (cm), Ht = H at time t, Ho = H at time t = 0, t = time (in seconds) since the first measurement of H as Ho, and F is a factor depending on the geometry of the hole:

    F = 4000 r / h' (20+D/ r)(2− h' /D)

where: r = radius of the cylindrical hole (cm), h' is the average depth of the water level in the hole relative to the water table in the soil (cm), found as h' =(Ho+Ht)/2, and D is the depth of the bottom of the hole relative to the water table in the soil (cm).

The picture shows a large variation of K-values measured with the augerhole method in an area of 100 ha. The ratio between the highest and lowest values is 25. The cumulative frequency distribution is lognormal and was made with the CumFreq program.

Related Magnitudes

Transmissivity

The transmissivity is a measure of how much water can be transmitted horizontally, such as to a pumping well.

> *Transmissivity* should not be confused with the similar word transmittance used in optics, meaning the fraction of incident light that passes through a sample.

An aquifer may consist of n soil layers. The transmissivity for horizontal flow T_i of the $i - th$ soil layer with a *saturated* thickness d_i and horizontal hydraulic conductivity K_i is:

$$T_i = K_i d_i$$

Transmissivity is directly proportional to horizontal hydraulic conductivity K_i and thickness d_i. Expressing K_i in m/day and d_i in m, the transmissivity T_i is found in units m²/day. The total transmissivity T_t of the aquifer is:

$$T_t = \sum T_i \text{ where } \sum \text{ signifies the summation over all layers } i = 1, 2, 3, \cdots, n..$$

The *apparent* horizontal hydraulic conductivity K_A of the aquifer is:

$$K_A = T_t / D_t$$

where D_t, the total thickness of the aquifer, is $D_t = \sum d_i$, with $i = 1, 2, 3, \cdots, n..$

The transmissivity of an aquifer can be determined from pumping tests.

Influence of the water table When a soil layer is above the water table, it is not saturated and does not contribute to the transmissivity. When the soil layer is entirely below the water table, its saturated thickness corresponds to the thickness of the soil layer itself. When the water table is inside a soil layer, the saturated thickness corresponds to the distance of the water table to the bottom of the layer. As the water table may behave dynamically, this thickness may change from place to place or from time to time, so that the transmissivity may vary accordingly. In a semi-confined aquifer, the water table is found within a soil layer with a negligibly small transmissivity, so that changes of the total transmissivity (Dt) resulting from changes in the level of the water table are negligibly small. When pumping water from an unconfined aquifer, where the water table is inside a soil layer with a significant transmissivity, the water table may be drawn down whereby the transmissivity reduces and the flow of water to the well diminishes.

Resistance

The *resistance* to vertical flow (R_i) of the $i-th$ soil layer with a *saturated* thickness d_i and vertical hydraulic conductivity Kv_i is:

$$R_i = d_i / Kv_i$$

Expressing Kv_i in m/day and d_i in m, the resistance (R_i) is expressed in days. The total resistance (Rt) of the aquifer is:

$$Rt = \Sigma R_i = \Sigma d_i / Kv_i$$

where Σ signifies the summation over all layers: $i = 1, 2, 3, \ldots n$

The *apparent* vertical hydraulic conductivity (Kv_A) of the aquifer is:

$$Kv_A = Dt / Rt$$

where Dt is the total thickness of the aquifer: $Dt = \Sigma d_i$, with $i = 1, 2, 3, \ldots n$

The resistance plays a role in aquifers where a sequence of layers occurs with varying horizontal permeability so that horizontal flow is found mainly in the layers with high horizontal permeability while the layers with low horizontal permeability transmit the water mainly in a vertical sense.

Anisotropy

When the horizontal and vertical hydraulic conductivity (Kh_i and Kv_i) of the $i-th$ soil layer differ considerably, the layer is said to be anisotropic with respect to hydraulic conductivity. When the *apparent* horizontal and vertical hydraulic conductivity (Kh_A and Kv_A) differ considerably, the aquifer is said to be anisotropic with respect to hydraulic conductivity. An aquifer is called *semi-confined* when a saturated layer with a relatively small horizontal hydraulic conductivity (the semi-confining layer or aquitard) overlies a layer with a relatively high horizontal hydraulic conductivity so that the flow of groundwater in the first layer is mainly vertical and in the second layer mainly horizontal. The resistance of a semi-confining top layer of an aquifer can be determined from pumping tests. When calculating flow to drains or to a well field in an aquifer with the aim to control the water table, the anisotropy is to be taken into account, otherwise the result may be erroneous.

Relative Properties

Because of their high porosity and permeability, sand and gravel aquifers have higher hydraulic conductivity than clay or unfractured granite aquifers. Sand or gravel aquifers would thus be easier to extract water from (e.g., using a pumping well) because of their high transmissivity, compared to clay or unfractured bedrock aquifers.

Hydraulic conductivity has units with dimensions of length per time (e.g., m/s, ft/day and (gal/day)/ft^2); transmissivity then has units with dimensions of length squared per time. The following table gives some typical ranges (illustrating the many orders of magnitude which are likely) for K values.

Hydraulic conductivity (K) is one of the most complex and important of the properties of aquifers in hydrogeology as the values found in nature:

- range over many orders of magnitude (the distribution is often considered to be lognormal),

- vary a large amount through space (sometimes considered to be randomly spatially distributed, or stochastic in nature),

- are directional (in general K is a symmetric second-rank tensor; e.g., vertical K values can be several orders of magnitude smaller than horizontal K values),

- are scale dependent (testing a m³ of aquifer will generally produce different results than a similar test on only a cm³ sample of the same aquifer),

- must be determined indirectly through field pumping tests, laboratory column flow tests or inverse computer simulation, (sometimes also from grain size analyses), and

- are very dependent (in a non-linear way) on the water content, which makes solving the unsaturated flow equation difficult. In fact, the variably saturated K for a single material varies over a wider range than the saturated K values for all types of materials (see chart below for an illustrative range of the latter).

Ranges of Values for Natural Materials

Table of Saturated Hydraulic Conductivity (K) Values Found in Nature

Values are for typical fresh groundwater conditions — using standard values of viscosity and specific gravity for water at 20°C and 1 atm. See the similar table derived from the same source for intrinsic permeability values.

K (cm/s)	10^2	10^1	$10^0=1$	10^{-1}	10^{-2}	10^{-3}	10^{-4}	10^{-5}	10^{-6}	10^{-7}	10^{-8}	10^{-9}	10^{-10}
K (ft/day)	10^5	10,000	1,000	100	10	1	0.1	0.01	0.001	0.0001	10^{-5}	10^{-6}	10^{-7}
Relative Permeability	Pervious						Semi-Pervious			Impervious			
Aquifer	Good						Poor			None			
Unconsolidated Sand & Gravel	Well Sorted Gravel		Well Sorted Sand or Sand & Gravel			Very Fine Sand, Silt, Loess, Loam							
Unconsolidated Clay & Organic					Peat		Layered Clay			Fat / Unweathered Clay			
Consolidated Rocks	Highly Fractured Rocks				Oil Reservoir Rocks			Fresh Sandstone		Fresh Limestone, Dolomite		Fresh Granite	

Source: modified from Bear, 1972

Evaporation

Aerosol of microscopic water droplets suspended in the air above a hot tea cup after that water vapor has sufficiently cooled and condensed. Water vapor is an invisible gas, but the clouds of condensed water droplets refract and disperse the sun light and so are visible.

Demonstration of evaporative cooling. When the sensor is dipped in ethanol and then taken out to evaporate, the instrument shows progressively lower temperature as the ethanol evaporates. Performed by Prof. Oliver Zajkov at the Physics Institute at the Ss. Cyril and Methodius University of Skopje, Macedonia.

Evaporation is a type of vaporization of a liquid that occurs from the surface of a liquid into a gaseous phase that is not saturated with the evaporating substance. The other type of vaporization is boiling, which is characterized by bubbles of saturated vapor forming in the liquid phase. Steam produced in a boiler is another example of evaporation occurring in a saturated vapor phase. Evaporation that occurs directly from the solid phase below the melting point, as commonly observed with ice at or below freezing or moth crystals (napthalene or paradichlorobenzene), is called sublimation.

On average, a fraction of the molecules in a glass of water have enough heat energy to escape from the liquid. Water molecules from the air enter the water in the glass, but as long as the relative humidity of the air in contact is less than 100% (saturation), the net transfer of water molecules will be to the air. The water in the glass will be cooled by the evaporation until an equilibrium is reached where the air supplies the amount of heat removed by the evaporating water. In an enclosed environment the water would evaporate until the air is saturated.

With sufficient temperature, the liquid would turn into vapor quickly. When the molecules collide, they transfer energy to each other in varying degrees, based on how they collide. Sometimes the transfer is so one-sided for a molecule near the surface that it ends up with enough energy to 'escape'.

Evaporation is an essential part of the water cycle. The sun (solar energy) drives evaporation of water from oceans, lakes, moisture in the soil, and other sources of water. In hydrology, evaporation and transpiration (which involves evaporation within plant stomata) are collectively termed evapotranspiration. Evaporation of water occurs when the surface of the liquid is exposed, allowing molecules to escape and form water vapor; this vapor can then rise up and form clouds. The tracking of evaporation from its source on the surface of the earth, through the atmosphere as vapor or clouds, and to its fate as precipitation closes the atmospheric water cycle, and embodies the concept of the precipitationshed.

Theory

For molecules of a liquid to evaporate, they must be located near the surface, they have to be moving in the proper direction, and have sufficient kinetic energy to overcome liquid-phase intermolecular forces. When only a small proportion of the molecules meet these criteria, the rate of evaporation is low. Since the kinetic energy of a molecule is proportional to its temperature, evaporation proceeds more quickly at higher temperatures. As the faster-moving molecules escape, the remaining molecules have lower average kinetic energy, and the temperature of the liquid decreases. This phenomenon is also called evaporative cooling. This is why evaporating sweat cools the human body. Evaporation also tends to proceed more quickly with higher flow rates between the gaseous and liquid phase and in liquids with higher vapor pressure. For example, laundry on a clothes line will dry (by evaporation) more rapidly on a windy day than on a still day. Three key parts to evaporation are heat, atmospheric pressure (determines the percent humidity) and air movement.

On a molecular level, there is no strict boundary between the liquid state and the vapor state. Instead, there is a Knudsen layer, where the phase is undetermined. Because this layer is only a few molecules thick, at a macroscopic scale a clear phase transition interface cannot be seen.

Liquids that do not evaporate visibly at a given temperature in a given gas (e.g., cooking oil at room temperature) have molecules that do not tend to transfer energy to each other in a pattern sufficient to frequently give a molecule the heat energy necessary to turn into vapor. However, these liquids *are* evaporating. It is just that the process is much slower and thus significantly less visible.

Evaporative Equilibrium

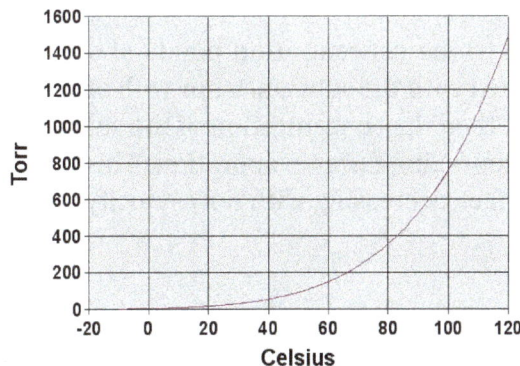

Vapor pressure of water vs. temperature. 760 Torr = 1 atm.

If evaporation takes place in an enclosed area, the escaping molecules accumulate as a vapor above the liquid. Many of the molecules return to the liquid, with returning molecules becoming more frequent as the density and pressure of the vapor increases. When the process of escape and return reaches an equilibrium, the vapor is said to be "saturated", and no further change in either vapor pressure and density or liquid temperature will occur. For a system consisting of vapor and liquid of a pure substance, this equilibrium state is directly related to the vapor pressure of the substance, as given by the Clausius–Clapeyron relation:

$$\ln\left(\frac{P_2}{P_1}\right) = -\frac{\Delta H_{vap}}{R}\left(\frac{1}{T_2} - \frac{1}{T_1}\right)$$

where P_1, P_2 are the vapor pressures at temperatures T_1, T_2 respectively, ΔH_{vap} is the enthalpy of vaporization, and R is the universal gas constant. The rate of evaporation in an open system is related to the vapor pressure found in a closed system. If a liquid is heated, when the vapor pressure reaches the ambient pressure the liquid will boil.

The ability for a molecule of a liquid to evaporate is based largely on the amount of kinetic energy an individual particle may possess. Even at lower temperatures, individual molecules of a liquid can evaporate if they have more than the minimum amount of kinetic energy required for vaporization.

Factors Influencing the Rate of Evaporation

Note: Air used here is a common example; however, the vapor phase can be other gases.

Concentration of the substance evaporating in the air

> If the air already has a high concentration of the substance evaporating, then the given substance will evaporate more slowly.

Concentration of other substances in the air

> If the air is already saturated with other substances, it can have a lower capacity for the substance evaporating.

Flow rate of air

> This is in part related to the concentration points above. If "fresh" air (i.e., air which is neither already saturated with the substance nor with other substances) is moving over the substance all the time, then the concentration of the substance in the air is less likely to go up with time, thus encouraging faster evaporation. This is the result of the boundary layer at the evaporation surface decreasing with flow velocity, decreasing the diffusion distance in the stagnant layer.

Inter-molecular forces

> The stronger the forces keeping the molecules together in the liquid state, the more energy one must get to escape. This is characterized by the enthalpy of vaporization.

Pressure

> Evaporation happens faster if there is less exertion on the surface keeping the molecules from launching themselves.

Surface area

A substance that has a larger surface area will evaporate faster, as there are more surface molecules per unit of volume that are potentially able to escape.

Temperature of the substance

the higher the temperature of the substance the greater the kinetic energy of the molecules at its surface and therefore the faster the rate of their evaporation.

In the US, the National Weather Service measures the actual rate of evaporation from a standardized "pan" open water surface outdoors, at various locations nationwide. Others do likewise around the world. The US data is collected and compiled into an annual evaporation map. The measurements range from under 30 to over 120 inches (3,000 mm) per year.

Thermodynamics

Evaporation is an endothermic process, in that heat is absorbed during evaporation.

Applications

- Industrial applications include many printing and coating processes; recovering salts from solutions; and drying a variety of materials such as lumber, paper, cloth and chemicals.

- The use of evaporation to dry or concentrate samples is a common preparatory step for many laboratory analyses such as spectroscopy and chromatography. Systems used for this purpose include rotary evaporators and centrifugal evaporators.

- When clothes are hung on a laundry line, even though the ambient temperature is below the boiling point of water, water evaporates. This is accelerated by factors such as low humidity, heat (from the sun), and wind. In a clothes dryer, hot air is blown through the clothes, allowing water to evaporate very rapidly.

- The Matki/Matka, a traditional Indian porous clay container used for storing and cooling water and other liquids.

- The botijo, a traditional Spanish porous clay container designed to cool the contained water by evaporation.

- Evaporative coolers, which can significantly cool a building by simply blowing dry air over a filter saturated with water.

Combustion Vaporization

Fuel droplets vaporize as they receive heat by mixing with the hot gases in the combustion chamber. Heat (energy) can also be received by radiation from any hot refractory wall of the combustion chamber.

Pre-combustion Vaporization

Internal combustion engines rely upon the vaporization of the fuel in the cylinders to form a fuel/

air mixture in order to burn well. The chemically correct air/fuel mixture for total burning of gasoline has been determined to be 15 parts air to one part gasoline or 15/1 by weight. Changing this to a volume ratio yields 8000 parts air to one part gasoline or 8,000/1 by volume.

Film Deposition

Thin films may be deposited by evaporating a substance and condensing it onto a substrate, or by dissolving the substance in a solvent, spreading the resulting solution thinly over a substrate, and evaporating the solvent.

Moisture Recycling

In hydrology, moisture recycling or precipitation recycling refer to the process by which a portion of the precipitated water that evapotranspired from a given area contributes to the precipitation over the same area. Moisture recycling is thus a component of the hydrologic cycle. The ratio of the locally derived precipitation (P_L) to total precipitation (P) is known as the recycling ratio $\rho : \rho = P_L / P$.

The recycling ratio is a diagnostic measure of the potential for interactions between land surface hydrology and regional climate. Land use changes, such as deforestation or agricultural intensification, have the potential to change the amount of precipitation that falls in a region. The recycling ratio for the entire world is one, and for a single point is zero. Estimates for the recycling ratio for the Amazon basin range from 24% to 56%, and for the Mississippi basin from 21% to 24%.

The concept of moisture recycling has been integrated into the concept of the precipitationshed. A precipitationshed is the upwind ocean and land surface that contributes evaporation to a given, downwind location's precipitation. In much the same way that a watershed is defined by a topographically explicit area that provides surface runoff, the precipitationshed is a probabilistically defined area within which evaporation, traveling via moisture recycling, provides precipitation for a specific point.

The American Institute of Biological Sciences published a paper in support of this concept in 2009. It also has been proposed, in the journal Atmospheric Chemistry and Physics, that evaporation rates from forested areas may exceed that of the oceans, creating zones of low pressure, which enhance the development of storms and rainfall through atmospheric moisture recycling.

Water Pollution

Water pollution is the contamination of water bodies (e.g. lakes, rivers, oceans, aquifers and groundwater). This form of environmental degradation occurs when pollutants are directly or indirectly discharged into water bodies without adequate treatment to remove harmful compounds.

Water pollution affects the entire biosphere – plants and organisms living in these bodies of water. In almost all cases the effect is damaging not only to individual species and population, but also to the natural biological communities.

Raw sewage and industrial waste in the New River as it passes from Mexicali to Calexico, California

Introduction

Pollution in the Lachine Canal, Canada

Water pollution is a major global problem which requires ongoing evaluation and revision of water resource policy at all levels (international down to individual aquifers and wells). It has been suggested that water pollution is the leading worldwide cause of deaths and diseases, and that it accounts for the deaths of more than 14,000 people daily. An estimated 580 people in India die of water pollution related illness every day. About 90 percent of the water in the cities of China is polluted. As of 2007, half a billion Chinese had no access to safe drinking water. In addition to the acute problems of water pollution in developing countries, developed countries also continue to struggle with pollution problems. For example, in the most recent national report on water quality in the United States, 44 percent of assessed stream miles, 64 percent of assessed lake acres, and 30 percent of assessed bays and estuarine square miles were classified as polluted. The head of China's national development agency said in 2007 that one quarter the length of China's seven main rivers were so poisoned the water harmed the skin.

Water is typically referred to as polluted when it is impaired by anthropogenic contaminants and either does not support a human use, such as drinking water, or undergoes a marked shift in its ability to support its constituent biotic communities, such as fish. Natural phenomena such as volcanoes, algae blooms, storms, and earthquakes also cause major changes in water quality and the ecological status of water.

Categories

Although interrelated, surface water and groundwater have often been studied and managed as separate resources. Surface water seeps through the soil and becomes groundwater. Conversely, groundwater can also feed surface water sources. Sources of surface water pollution are generally grouped into two categories based on their origin.

Point Sources

Point source pollution – Shipyard – Rio de Janeiro.

Point source water pollution refers to contaminants that enter a waterway from a single, identifiable source, such as a pipe or ditch. Examples of sources in this category include discharges from a sewage treatment plant, a factory, or a city storm drain. The U.S. Clean Water Act (CWA) defines point source for regulatory enforcement purposes. The CWA definition of point source was amended in 1987 to include municipal storm sewer systems, as well as industrial storm water, such as from construction sites.

Non-point Sources

Nonpoint source pollution refers to diffuse contamination that does not originate from a single discrete source. NPS pollution is often the cumulative effect of small amounts of contaminants gathered from a large area. A common example is the leaching out of nitrogen compounds from fertilized agricultural lands. Nutrient runoff in storm water from "sheet flow" over an agricultural field or a forest are also cited as examples of NPS pollution.

Contaminated storm water washed off of parking lots, roads and highways, called urban runoff, is sometimes included under the category of NPS pollution. However, because this runoff is typically channeled into storm drain systems and discharged through pipes to local surface waters, it becomes a point source.

Blue drain and yellow fish symbol used by the UK Environment Agency to raise awareness of the ecological impacts of contaminating surface drainage

Groundwater Pollution

Interactions between groundwater and surface water are complex. Consequently, groundwater pollution, also referred to as groundwater contamination, is not as easily classified as surface water pollution. By its very nature, groundwater aquifers are susceptible to contamination from sources that may not directly affect surface water bodies, and the distinction of point vs. non-point source may be irrelevant. A spill or ongoing release of chemical or radionuclide contaminants into soil (located away from a surface water body) may not create point or non-point source pollution but can contaminate the aquifer below, creating a toxic plume. The movement of the plume, called a plume front, may be analyzed through a hydrological transport model or groundwater model. Analysis of groundwater contamination may focus on soil characteristics and site geology, hydrogeology, hydrology, and the nature of the contaminants.

Causes

The specific contaminants leading to pollution in water include a wide spectrum of chemicals, pathogens, and physical changes such as elevated temperature and discoloration. While many of the chemicals and substances that are regulated may be naturally occurring (calcium, sodium, iron, manganese, etc.) the concentration is often the key in determining what is a natural component of water and what is a contaminant. High concentrations of naturally occurring substances can have negative impacts on aquatic flora and fauna.

Oxygen-depleting substances may be natural materials such as plant matter (e.g. leaves and grass) as well as man-made chemicals. Other natural and anthropogenic substances may cause turbidity (cloudiness) which blocks light and disrupts plant growth, and clogs the gills of some fish species.

Many of the chemical substances are toxic. Pathogens can produce waterborne diseases in either human or animal hosts. Alteration of water's physical chemistry includes acidity (change in pH), electrical conductivity, temperature, and eutrophication. Eutrophication is an increase in the concentration of chemical nutrients in an ecosystem to an extent that increases in the primary productivity of the ecosystem. Depending on the degree of eutrophication, subsequent negative environmental effects such as anoxia (oxygen depletion) and severe reductions in water quality may occur, affecting fish and other animal populations.

Pathogens

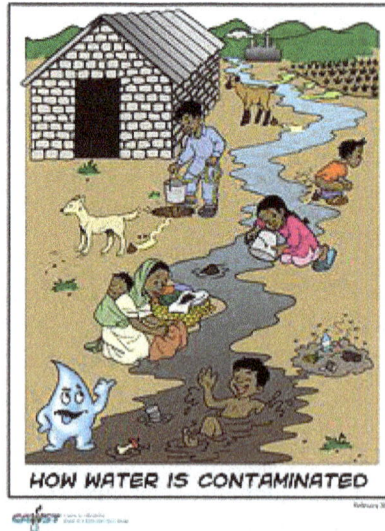

Poster to teach people in South Asia about human activities leading to the pollution of water sources

A manhole cover unable to contain a sanitary sewer overflow.

Fecal sludge collected from pit latrines is dumped into a river at the Korogocho slum in Nairobi, Kenya.

Disease-causing microorganisms are referred to as pathogens. Although the vast majority of bacteria are either harmless or beneficial, a few pathogenic bacteria can cause disease. Coliform bacteria, which are not an actual cause of disease, are commonly used as a bacterial indicator of water pollution. Other microorganisms sometimes found in surface waters that have caused human health problems include:

- *Burkholderia pseudomallei*

- *Cryptosporidium parvum*

- *Giardia lamblia*

- *Salmonella*

- *Norovirus* and other viruses

- *Parasitic worms including the Schistosoma type*

High levels of pathogens may result from on-site sanitation systems (septic tanks, pit latrines) or inadequately treated sewage discharges. This can be caused by a sewage plant designed with less than secondary treatment (more typical in less-developed countries). In developed countries, older cities with aging infrastructure may have leaky sewage collection systems (pipes, pumps, valves), which can cause sanitary sewer overflows. Some cities also have combined sewers, which may discharge untreated sewage during rain storms.

Muddy river polluted by sediment.

Pathogen discharges may also be caused by poorly managed livestock operations.

Organic, Inorganic and Macroscopic Contaminants

Contaminants may include organic and inorganic substances.

A garbage collection boom in an urban-area stream in Auckland, New Zealand.

Organic water pollutants include:

- Detergents

- Disinfection by-products found in chemically disinfected drinking water, such as chloroform

- Food processing waste, which can include oxygen-demanding substances, fats and grease

- Insecticides and herbicides, a huge range of organohalides and other chemical compounds

- Petroleum hydrocarbons, including fuels (gasoline, diesel fuel, jet fuels, and fuel oil) and lubricants (motor oil), and fuel combustion byproducts, from storm water runoff

- Volatile organic compounds, such as industrial solvents, from improper storage.

- Chlorinated solvents, which are dense non-aqueous phase liquids, may fall to the bottom of reservoirs, since they don't mix well with water and are denser.

    o Polychlorinated biphenyl (PCBs)

    o Trichloroethylene

- Perchlorate

- Various chemical compounds found in personal hygiene and cosmetic products

- Drug pollution involving pharmaceutical drugs and their metabolites

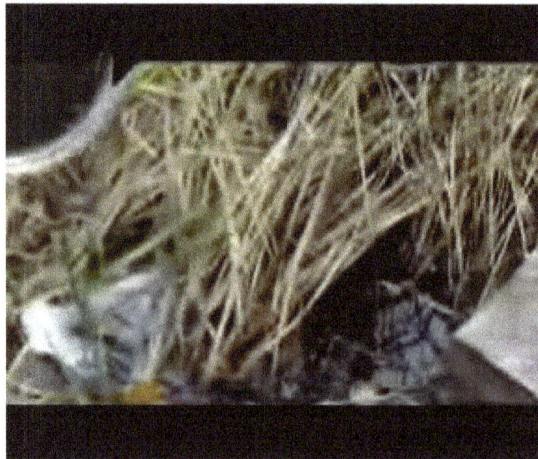

Macroscopic Pollution in Parks Milwaukee, WI

Inorganic water pollutants include:

- Acidity caused by industrial discharges (especially sulfur dioxide from power plants)

- Ammonia from food processing waste

- Chemical waste as industrial by-products

- Fertilizers containing nutrients--nitrates and phosphates—which are found in storm water runoff from agriculture, as well as commercial and residential use

- Heavy metals from motor vehicles (via urban storm water runoff) and acid mine drainage

- Silt (sediment) in runoff from construction sites, logging, slash and burn practices or land clearing sites.

Macroscopic pollution – large visible items polluting the water – may be termed "floatables" in an urban storm water context, or marine debris when found on the open seas, and can include such items as:

- Trash or garbage (e.g. paper, plastic, or food waste) discarded by people on the ground, along with accidental or intentional dumping of rubbish, that are washed by rainfall into storm drains and eventually discharged into surface waters

- Nurdles, small ubiquitous waterborne plastic pellets

- Shipwrecks, large derelict ships.

The Brayton Point Power Station in Massachusetts discharges heated water to Mount Hope Bay.

Thermal pollution

Thermal pollution is the rise or fall in the temperature of a natural body of water caused by human influence. Thermal pollution, unlike chemical pollution, results in a change in the physical properties of water. A common cause of thermal pollution is the use of water as a coolant by power plants and industrial manufacturers. Elevated water temperatures decrease oxygen levels, which can kill fish and alter food chain composition, reduce species biodiversity, and foster invasion by new thermophilic species. Urban runoff may also elevate temperature in surface waters.

Thermal pollution can also be caused by the release of very cold water from the base of reservoirs into warmer rivers.

Transport and Chemical Reactions of Water Pollutants

Most water pollutants are eventually carried by rivers into the oceans. In some areas of the world the influence can be traced one hundred miles from the mouth by studies using hydrology transport models. Advanced computer models such as SWMM or the DSSAM Model have been used in many locations worldwide to examine the fate of pollutants in aquatic systems. Indicator filter feeding species such as copepods have also been used to study pollutant fates in the New York Bight, for example. The highest toxin loads are not directly at the mouth of the Hudson River, but 100 km (62 mi) south, since several days are required for incorporation into planktonic tissue. The Hudson discharge flows south along the coast due to the coriolis force. Further south are areas of oxygen depletion caused by chemicals using up oxygen and by algae blooms, caused by excess nutrients from algal cell death and decomposition. Fish and shellfish kills have been reported, because toxins climb the food chain after small fish consume copepods, then large fish eat smaller fish, etc. Each successive step up the food chain causes a cumulative concentration of pollutants such as heavy metals (e.g. mercury) and persistent organic pollutants such as DDT. This is known as bio-magnification, which is occasionally used interchangeably with bio-accumulation.

A polluted river draining an abandoned copper mine on Anglesey

Large gyres (vortexes) in the oceans trap floating plastic debris. The North Pacific Gyre, for example, has collected the so-called "Great Pacific Garbage Patch", which is now estimated to be one hundred times the size of Texas. Plastic debris can absorb toxic chemicals from ocean pollution, potentially poisoning any creature that eats it. Many of these long-lasting pieces wind up in the stomachs of marine birds and animals. This results in obstruction of digestive pathways, which leads to reduced appetite or even starvation.

Many chemicals undergo reactive decay or chemical change, especially over long periods of time in groundwater reservoirs. A noteworthy class of such chemicals is the chlorinated hydrocarbons such as trichloroethylene (used in industrial metal degreasing and electronics manufacturing) and tetrachloroethylene used in the dry cleaning industry. Both of these chemicals, which are carcinogens themselves, undergo partial decomposition reactions, leading to new hazardous chemicals (including dichloroethylene and vinyl chloride).

Groundwater pollution is much more difficult to abate than surface pollution because groundwater can move great distances through unseen aquifers. Non-porous aquifers such as clays partially purify water of bacteria by simple filtration (adsorption and absorption), dilution, and, in some cases, chemical reactions and biological activity; however, in some cases, the pollutants merely transform to soil contaminants. Groundwater that moves through open fractures and caverns is not filtered and can be transported as easily as surface water. In fact, this can be aggravated by the human tendency to use natural sinkholes as dumps in areas of karst topography.

There are a variety of secondary effects stemming not from the original pollutant, but a derivative condition. An example is silt-bearing surface runoff, which can inhibit the penetration of sunlight through the water column, hampering photosynthesis in aquatic plants.

Measurement

Environmental scientists preparing water autosamplers.

Water pollution may be analyzed through several broad categories of methods: physical, chemical and biological. Most involve collection of samples, followed by specialized analytical tests. Some methods may be conducted *in situ*, without sampling, such as temperature. Government agencies and research organizations have published standardized, validated analytical test methods to facilitate the comparability of results from disparate testing events.

Sampling

Sampling of water for physical or chemical testing can be done by several methods, depending on the accuracy needed and the characteristics of the contaminant. Many contamination events are sharply restricted in time, most commonly in association with rain events. For this reason "grab" samples are often inadequate for fully quantifying contaminant levels. Scientists gathering this type of data often employ auto-sampler devices that pump increments of water at either time or discharge intervals.

Sampling for biological testing involves collection of plants and/or animals from the surface water body. Depending on the type of assessment, the organisms may be identified for biosurveys (population counts) and returned to the water body, or they may be dissected for bioassays to determine toxicity.

Physical Testing

Common physical tests of water include temperature, solids concentrations (e.g., total suspended solids (TSS)) and turbidity.

Chemical Testing

Water samples may be examined using the principles of analytical chemistry. Many published test methods are available for both organic and inorganic compounds. Frequently used methods include pH, biochemical oxygen demand (BOD), chemical oxygen demand (COD), nutrients (nitrate and phosphorus compounds), metals (including copper, zinc, cadmium, lead and mercury), oil and grease, total petroleum hydrocarbons (TPH), and pesticides.

Biological Testing

Biological testing involves the use of plant, animal, and/or microbial indicators to monitor the health of an aquatic ecosystem. They are any biological species or group of species whose function, population, or status can reveal what degree of ecosystem or environmental integrity is present. One example of a group of bio-indicators are the copepods and other small water crustaceans that are present in many water bodies. Such organisms can be monitored for changes (biochemical, physiological, or behavioral) that may indicate a problem within their ecosystem.

Control of Pollution

Decisions on the type and degree of treatment and control of wastes, and the disposal and use of adequately treated wastewater, must be based on a consideration all the technical factors of each drainage basin, in order to prevent any further contamination or harm to the environment.

Sewage Treatment

Deer Island Wastewater Treatment Plant serving Boston, Massachusetts and vicinity.

In urban areas of developed countries, domestic sewage is typically treated by centralized sewage treatment plants. Well-designed and operated systems (i.e., secondary treatment or better) can remove 90 percent or more of the pollutant load in sewage. Some plants have additional systems to remove nutrients and pathogens.

Cities with sanitary sewer overflows or combined sewer overflows employ one or more engineering approaches to reduce discharges of untreated sewage, including:

- utilizing a green infrastructure approach to improve storm water management capacity throughout the system, and reduce the hydraulic overloading of the treatment plant

- repair and replacement of leaking and malfunctioning equipment

- increasing overall hydraulic capacity of the sewage collection system (often a very expensive option).

A household or business not served by a municipal treatment plant may have an individual septic tank, which pre-treats the wastewater on site and infiltrates it into the soil.

Industrial Wastewater Treatment

Dissolved air flotation system for treating industrial wastewater.

Some industrial facilities generate ordinary domestic sewage that can be treated by municipal facilities. Industries that generate wastewater with high concentrations of conventional pollutants (e.g. oil and grease), toxic pollutants (e.g. heavy metals, volatile organic compounds) or other non-conventional pollutants such as ammonia, need specialized treatment systems. Some of these facilities can install a pre-treatment system to remove the toxic components, and then send the partially treated wastewater to the municipal system. Industries generating large volumes of wastewater typically operate their own complete on-site treatment systems. Some industries have been successful at redesigning their manufacturing processes to reduce or eliminate pollutants, through a process called pollution prevention.

Heated water generated by power plants or manufacturing plants may be controlled with:

- cooling ponds, man-made bodies of water designed for cooling by evaporation, convection, and radiation

- cooling towers, which transfer waste heat to the atmosphere through evaporation and/or heat transfer

- cogeneration, a process where waste heat is recycled for domestic and/or industrial heating purposes.

Riparian buffer lining a creek in Iowa.

Agricultural Wastewater Treatment

Non point source controls Sediment (loose soil) washed off fields is the largest source of agricultural pollution in the United States. Farmers may utilize erosion controls to reduce runoff flows and retain soil on their fields. Common techniques include contour plowing, crop mulching, crop rotation, planting perennial crops and installing riparian buffers.

Nutrients (nitrogen and phosphorus) are typically applied to farmland as commercial fertilizer, animal manure, or spraying of municipal or industrial wastewater (effluent) or sludge. Nutrients may also enter runoff from crop residues, irrigation water, wildlife, and atmospheric deposition. Farmers can develop and implement nutrient management plans to reduce excess application of nutrients and reduce the potential for nutrient pollution.

To minimize pesticide impacts, farmers may use Integrated Pest Management (IPM) techniques (which can include biological pest control) to maintain control over pests, reduce reliance on chemical pesticides, and protect water quality.

Feedlot in the United States

Point source wastewater treatment Farms with large livestock and poultry operations, such as factory farms, are called *concentrated animal feeding operations* or *feedlots* in the US and are being subject to increasing government regulation. Animal slurries are usually treated by containment in anaerobic lagoons before disposal by spray or trickle application to grassland. Constructed wetlands are sometimes used to facilitate treatment of animal wastes. Some animal slurries are treated by mixing with straw and composted at high temperature to produce a bacteriologically sterile and friable manure for soil improvement.

Erosion and Sediment Control from Construction Sites

Silt fence installed on a construction site.

Sediment from construction sites is managed by installation of:

- erosion controls, such as mulching and hydroseeding, and

- sediment controls, such as sediment basins and silt fences.

Discharge of toxic chemicals such as motor fuels and concrete washout is prevented by use of:

- spill prevention and control plans, and

- specially designed containers (e.g. for concrete washout) and structures such as overflow controls and diversion berms.

Control of Urban Runoff (Storm Water)

Retention basin for controlling urban runoff

Effective control of urban runoff involves reducing the velocity and flow of storm water, as well as reducing pollutant discharges. Local governments use a variety of storm water management techniques to reduce the effects of urban runoff. These techniques, called best management practices (BMPs) in the U.S., may focus on water quantity control, while others focus on improving water quality, and some perform both functions.

Pollution prevention practices include low-impact development techniques, installation of green roofs and improved chemical handling (e.g. management of motor fuels & oil, fertilizers and pesticides). Runoff mitigation systems include infiltration basins, bioretention systems, constructed wetlands, retention basins and similar devices.

Thermal pollution from runoff can be controlled by storm water management facilities that absorb the runoff or direct it into groundwater, such as bioretention systems and infiltration basins. Retention basins tend to be less effective at reducing temperature, as the water may be heated by the sun before being discharged to a receiving stream.

References

- Metzger, Bruce M.; Coogan, Michael D. (1993). The Oxford Companion to the Bible. New York, NY: Oxford University Press. p. 369. ISBN 0195046455.

- Merrill, Eugene H.; Rooker, Mark F.; Grisanti, Michael A. (2011). The World and the Word. Nashville, TN: B&H Academic. p. 430. ISBN 9780805440317.

- J.Boonstra and R.A.L.Kselik, SATEM 2002: Software for aquifer test evaluation, 2001. Publ. 57, International Institute for Land reclamation and Improvement (ILRI), Wageningen, The Netherlands. ISBN 90-70754-54-1

- Goel, P.K. (2006). Water Pollution - Causes, Effects and Control. New Delhi: New Age International. p. 179. ISBN 978-81-224-1839-2.

- Kennish, Michael J. (1992). Ecology of Estuaries: Anthropogenic Effects. Marine Science Series. Boca Raton, FL: CRC Press. pp. 415–17. ISBN 978-0-8493-8041-9.

- Laws, Edward A. (2000). Aquatic Pollution: An Introductory Text. New York: John Wiley and Sons. p. 430. ISBN 978-0-471-34875-7.

- For example, see Baird, Rodger B.; Clesceri, Leonore S.; Eaton, Andrew D.; et al., eds. (2012). Standard Methods for the Examination of Water and Wastewater (22nd ed.). Washington, DC: American Public Health Association. ISBN 978-0875530130.

- G. Allen Burton, Jr., Robert Pitt (2001). Stormwater Effects Handbook: A Toolbox for Watershed Managers, Scientists, and Engineers. New York: CRC/Lewis Publishers. ISBN 0-87371-924-7.

- Gleeson, Tom; Wada, Yoshihide; Bierkens, Marc F. P.; van Beek, Ludovicus P. H. (9 August 2012). "Water balance of global aquifers revealed by groundwater footprint". Nature. 488 (7410): 197–200. doi:10.1038/nature11295. Retrieved 2013-05-29.

- Justin Gillis (April 26, 2012). "Study Indicates a Greater Threat of Extreme Weather". The New York Times. Retrieved April 27, 2012.

- Zaikab, Gwyneth Dickey (2011-03-28). "Marine microbes digest plastic". Nature. Macmillan. doi:10.1038/news.2011.191. ISSN 0028-0836.

9

Flood: A Hydrological Study

Floods occur when water inundates low lying areas which were not covered by water. They cause damage to both life and property. Floods can be caused due to the overflow of water bodies like lakes, rivers, streams etc. or by high precipitation. They can develop slowly or in a matter of minutes. This chapter comprehensively analyses floods, its types and related terminology like flood stage, flood warning, floodgate and flood barrier. The aspects elucidated in this chapter are of vital importance, and provide a better understanding of hydrology.

Flood

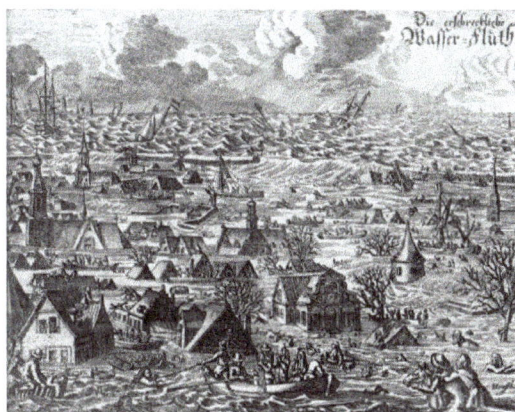

Contemporary picture of the flood that struck the North Sea coast of Germany and Denmark in October 1634.

A flood is an overflow of water that submerges land which is usually dry. The European Union (EU) Floods Directive defines a flood as a covering by water of land not normally covered by water. In the sense of "flowing water", the word may also be applied to the inflow of the tide.

Flooding may occur as an overflow of water from water bodies, such as a river, lake, or ocean, in which the water overtops or breaks levees, resulting in some of that water escaping its usual boundaries, or it may occur due to an accumulation of rainwater on saturated ground in an areal flood. While the size of a lake or other body of water will vary with seasonal changes in precipitation and snow melt, these changes in size are unlikely to be considered significant unless they flood property or drown domestic animals.

Floods can also occur in rivers when the flow rate exceeds the capacity of the river channel, particularly at bends or meanders in the waterway. Floods often cause damage to homes and businesses if they are in the natural flood plains of rivers. While riverine flood damage can be eliminated by moving away from rivers and other bodies of water, people have traditionally lived and worked by rivers because the land is usually flat and fertile and because rivers provide easy travel and access to commerce and industry.

Some floods develop slowly, while others such as flash floods, can develop in just a few minutes and without visible signs of rain. Additionally, floods can be local, impacting a neighborhood or community, or very large, affecting entire river basins.

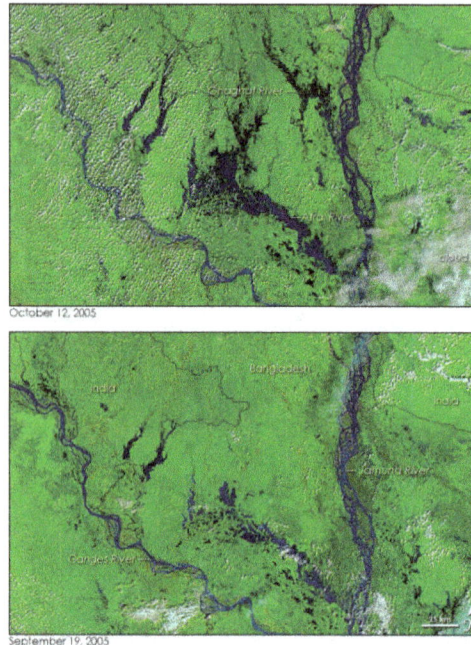

Dozens of villages were inundated when rain pushed the rivers of northwestern Bangladesh over their banks in early October 2005. The Moderate Resolution Imaging Spectroradiometer (MODIS) on NASA's Terra satellite captured the top image of the flooded Ghaghat and Atrai Rivers on October 12, 2005. The deep blue of the rivers is spread across the countryside in the flood image.

Etymology

The word "flood" comes from the Old English *flod*, a word common to Germanic languages (compare German *Flut*, Dutch *vloed* from the same root as is seen in *flow, float*; also compare with Latin *fluctus, flumen*). Deluge myths are mythical stories of a great flood sent by a deity or deities to destroy civilization as an act of divine retribution, and they are featured in the mythology of many cultures.

Principal Types

Areal

Floods can happen on flat or low-lying areas when water is supplied by rainfall or snowmelt more rapidly than it can either infiltrate or run off. The excess accumulates in place, sometimes to hazardous depths. Surface soil can become saturated, which effectively stops infiltration, where the water table is shallow, such as a floodplain, or from intense rain from one or a series of storms. Infiltration also is slow to negligible through frozen ground, rock, concrete, paving, or roofs. Areal flooding begins in flat areas like floodplains and in local depressions not connected to a stream channel, because the velocity of overland flow depends on the surface slope. Endorheic basins may experience areal flooding during periods when precipitation exceeds evaporation.

Riverine (Channel)

Floods occur in all types of river and stream channels, from the smallest ephemeral streams in humid zones to normally-dry channels in arid climates to the world's largest rivers. When overland flow occurs on tilled fields, it can result in a muddy flood where sediments are picked up by run off and carried as suspended matter or bed load. Localized flooding may be caused or exacerbated by drainage obstructions such as landslides, ice, debris, or beaver dams.

Slow-rising floods most commonly occur in large rivers with large catchment areas. The increase in flow may be the result of sustained rainfall, rapid snow melt, monsoons, or tropical cyclones. However, large rivers may have rapid flooding events in areas with dry climate, since they may have large basins but small river channels and rainfall can be very intense in smaller areas of those basins.

Rapid flooding events, including flash floods, more often occur on smaller rivers, rivers with steep valleys, rivers that flow for much of their length over impermeable terrain, or normally-dry channels. The cause may be localized convective precipitation (intense thunderstorms) or sudden release from an upstream impoundment created behind a dam, landslide, or glacier. In one instance, a flash flood killed 8 people enjoying the water on a Sunday afternoon at a popular waterfall in a narrow canyon. Without any observed rainfall, the flow rate increased from about 50 to 1,500 cubic feet per second (1.4 to 42 m³/s) in just one minute. Two larger floods occurred at the same site within a week, but no one was at the waterfall on those days. The deadly flood resulted from a thunderstorm over part of the drainage basin, where steep, bare rock slopes are common and the thin soil was already saturated.

Flash floods are the most common flood type in normally-dry channels in arid zones, known as arroyos in the southwest United States and many other names elsewhere. In that setting, the first flood water to arrive is depleted as it wets the sandy stream bed. The leading edge of the flood thus advances more slowly than later and higher flows. As a result, the rising limb of the hydrograph becomes ever quicker as the flood moves downstream, until the flow rate is so great that the depletion by wetting soil becomes insignificant.

Estuarine and Coastal

Flooding in estuaries is commonly caused by a combination of sea tidal surges caused by winds and low barometric pressure, and they may be exacerbated by high upstream river flow.

Coastal areas may be flooded by storm events at sea, resulting in waves over-topping defenses or in severe cases by tsunami or tropical cyclones. A storm surge, from either a tropical cyclone or an extratropical cyclone, falls within this category. Research from the NHC (National Hurricane Center) explains: "Storm surge is an abnormal rise of water generated by a storm, over and above the predicted astronomical tides. Storm surge should not be confused with storm tide, which is defined as the water level rise due to the combination of storm surge and the astronomical tide. This rise in water level can cause extreme flooding in coastal areas particularly when storm surge coincides with normal high tide, resulting in storm tides reaching up to 20 feet or more in some cases."

Urban Flooding

Urban flooding is the inundation of land or property in a built environment, particularly in more densely populated areas, caused by rainfall overwhelming the capacity of drainage systems, such as storm sewers. Although sometimes triggered by events such as flash flooding or snowmelt, urban flooding is a condition, characterized by its repetitive and systemic impacts on communities, that can happen regardless of whether or not affected communities are located within designated floodplains or near any body of water. Aside from potential overflow of rivers and lakes, snowmelt, stormwater or water released from damaged water mains may accumulate on property and in public rights-of-way, seep through building walls and floors, or backup into buildings through sewer pipes, toilets and sinks.

In urban areas, flood effects can be exacerbated by existing paved streets and roads, which increase the speed of flowing water.

The flood flow in urbanized areas constitutes a hazard to both the population and infrastructure. Some recent catastrophes include the inundations of Nîmes (France) in 1998 and Vaison-la-Romaine (France) in 1992, the flooding of New Orleans (USA) in 2005, and the flooding in Rockhampton, Bundaberg, Brisbane during the 2010–2011 summer in Queensland (Australia). Flood flows in urban environments have been studied relatively recently despite many centuries of flood events. Some recent research has considered the criteria for safe evacuation of individuals in flooded areas.

Catastrophic

Catastrophic riverine flooding is usually associated with major infrastructure failures such as the collapse of a dam, but they may also be caused by drainage channel modification from a landslide, earthquake or volcanic eruption. Examples include outburst floods and lahars. Tsunamis can cause catastrophic coastal flooding, most commonly resulting from undersea earthquakes.

Causes

Upslope Factors

The amount, location, and timing of water reaching a drainage channel from natural precipitation and controlled or uncontrolled reservoir releases determines the flow at downstream locations. Some precipitation evaporates, some slowly percolates through soil, some may be temporarily sequestered as snow or ice, and some may produce rapid runoff from surfaces including rock, pavement, roofs, and saturated or frozen ground. The fraction of incident precipitation promptly reaching a drainage channel has been observed from nil for light rain on dry, level ground to as high as 170 percent for warm rain on accumulated snow.

Most precipitation records are based on a measured depth of water received within a fixed time interval. *Frequency* of a precipitation threshold of interest may be determined from the number of measurements exceeding that threshold value within the total time period for which observations are available. Individual data points are converted to *intensity* by dividing each measured depth by the period of time between observations. This intensity will be less than the actual peak intensity if the *duration* of the rainfall event was less than the fixed time interval for which measurements are reported. Convective precipitation events (thunderstorms) tend to produce shorter duration storm events than orographic precipitation. Duration, intensity, and frequency of rainfall events are important to flood prediction. Short duration precipitation is more significant to flooding within small drainage basins.

The most important upslope factor in determining flood magnitude is the land area of the watershed upstream of the area of interest. Rainfall intensity is the second most important factor for watersheds of less than approximately 30 square miles or 80 square kilometres. The main channel slope is the second most important factor for larger watersheds. Channel slope and rainfall intensity become the third most important factors for small and large watersheds, respectively.

Time of Concentration is the time required for runoff from the most distant point of the upstream drainage area to reach the point of the drainage channel controlling flooding of the area of interest. The time of concentration defines the critical duration of peak rainfall for the area of interest. The critical duration of intense rainfall might be only a few minutes for roof and parking lot drainage structures, while cumulative rainfall over several days would be critical for river basins.

| ronoff erosion | rapid creek flooding | trapped woman on top of vehicle | runoff and filter soxx |

Downslope Factors

Water flowing downhill ultimately encounters downstream conditions slowing movement. The final limitation is often the ocean or a natural or artificial lake. Elevation changes such as tidal fluctuations are significant determinants of coastal and estuarine flooding. Less predictable events like tsunamis and storm surges may also cause elevation changes in large bodies of water. Elevation of flowing water is controlled by the geometry of the flow channel. Flow channel restrictions like bridges and canyons tend to control water elevation above the restriction. The actual control point for any given reach of the drainage may change with changing water elevation, so a closer point may control for lower water levels until a more distant point controls at higher water levels.

Effective flood channel geometry may be changed by growth of vegetation, accumulation of ice or debris, or construction of bridges, buildings, or levees within the flood channel.

Coincidence

Extreme flood events often result from coincidence such as unusually intense, warm rainfall melting heavy snow pack, producing channel obstructions from floating ice, and releasing small impoundments like beaver dams. Coincident events may cause extensive flooding to be more frequent than anticipated from simplistic statistical prediction models considering only precipitation runoff flowing within unobstructed drainage channels. Debris modification of channel geometry is common when heavy flows move uprooted woody vegetation and flood-damaged structures and vehicles, including boats and railway equipment. Recent field measurements during the 2010–2011 Queensland floods showed that any criterion solely based upon the flow velocity, water depth or specific momentum cannot account for the hazards caused by velocity and water depth fluctuations. These considerations ignore further the risks associated with large debris entrained by the flow motion.

Some researchers have mentioned the storage effect in urban areas with transportation corridors created by cut and fill. Culverted fills may be converted to impoundments if the culverts become blocked by debris, and flow may be diverted along streets. Several studies have looked into the flow patterns and redistribution in streets during storm events and the implication on flood modelling.

Effects

Primary Effects

The primary effects of flooding include loss of life, damage to buildings and other structures, including bridges, sewerage systems, roadways, and canals.

Floods also frequently damage power transmission and sometimes power generation, which then has knock-on effects caused by the loss of power. This includes loss of drinking water treatment and water supply, which may result in loss of drinking water or severe water contamination. It may also cause the loss of sewage disposal facilities. Lack of clean water combined with human sewage in the flood waters raises the risk of waterborne diseases, which can include typhoid, giardia, cryptosporidium, cholera and many other diseases depending upon the location of the flood.

Damage to roads and transport infrastructure may make it difficult to mobilize aid to those affected or to provide emergency health treatment.

Flood waters typically inundate farm land, making the land unworkable and preventing crops from being planted or harvested, which can lead to shortages of food both for humans and farm animals. Entire harvests for a country can be lost in extreme flood circumstances. Some tree species may not survive prolonged flooding of their root systems

Secondary and Long-term Effects

Economic hardship due to a temporary decline in tourism, rebuilding costs, or food shortages leading to price increases is a common after-effect of severe flooding. The impact on those affected may cause psychological damage to those affected, in particular where deaths, serious injuries and loss of property occur.

Urban flooding can lead to chronically wet houses, which are linked to an increase in respiratory problems and other illnesses. Urban flooding also has significant economic implications for affected neighborhoods. In the United States, industry experts estimate that wet basements can lower property values by 10-25 percent and are cited among the top reasons for not purchasing a home. According to the U.S. Federal Emergency Management Agency (FEMA), almost 40 percent of small businesses never reopen their doors following a flooding disaster. In the United States, insurance is available against flood damage to both homes and businesses.

Benefits

Floods (in particular more frequent or smaller floods) can also bring many benefits, such as recharging ground water, making soil more fertile and increasing nutrients in some soils. Flood waters provide much needed water resources in arid and semi-arid regions where precipitation can be very unevenly distributed throughout the year and kills pests in the farming land. Freshwater floods particularly play an important role in maintaining ecosystems in river corridors and are a key factor in maintaining floodplain biodiversity. Flooding can spread nutrients to lakes and rivers, which can lead to increased biomass and improved fisheries for a few years.

For some fish species, an inundated floodplain may form a highly suitable location for spawning with few predators and enhanced levels of nutrients or food. Fish, such as the weather fish, make use of floods in order to reach new habitats. Bird populations may also profit from the boost in food production caused by flooding.

Periodic flooding was essential to the well-being of ancient communities along the Tigris-Euphrates Rivers, the Nile River, the Indus River, the Ganges and the Yellow River among others. The viability of hydropower, a renewable source of energy, is also higher in flood prone regions.

Flood Safety Planning

At the most basic level, the best defense against floods is to seek higher ground for high-value uses while balancing the foreseeable risks with the benefits of occupying flood hazard zones. Critical community-safety facilities, such as hospitals, emergency-operations centers, and police, fire, and rescue services, should be built in areas least at risk of flooding. Structures, such as bridges, that must unavoidably be in flood hazard areas should be designed to withstand flooding. Areas most at risk for flooding could be put to valuable uses that could be abandoned temporarily as people retreat to safer areas when a flood is imminent.

Planning for flood safety involves many aspects of analysis and engineering, including:

- observation of previous and present flood heights and inundated areas,
- statistical, hydrologic, and hydraulic model analyses,
- mapping inundated areas and flood heights for future flood scenarios,
- long-term land use planning and regulation,
- engineering design and construction of structures to control or withstand flooding,
- intermediate-term monitoring, forecasting, and emergency-response planning, and
- short-term monitoring, warning, and response operations.

Each topic presents distinct yet related questions with varying scope and scale in time, space, and the people involved. Attempts to understand and manage the mechanisms at work in floodplains have been made for at least six millennia.

In the United States, the Association of State Floodplain Managers works to promote education, policies, and activities that mitigate current and future losses, costs, and human suffering caused by flooding and to protect the natural and beneficial functions of floodplains - all without causing adverse impacts. A portfolio of best practice examples for disaster mitigation in the United States is available from the Federal Emergency Management Agency.

Control

In many countries around the world, waterways prone to floods are often carefully managed. Defenses such as detention basins, levees, bunds, reservoirs, and weirs are used to prevent waterways from overflowing their banks. When these defenses fail, emergency measures such as sandbags or portable inflatable tubes are often used to try to stem flooding. Coastal flooding has been addressed in portions of Europe and the Americas with coastal defenses, such as sea walls, beach nourishment, and barrier islands.

In the riparian zone near rivers and streams, erosion control measures can be taken to try to slow down or reverse the natural forces that cause many waterways to meander over long periods of time. Flood controls, such as dams, can be built and maintained over time to try to reduce the occurrence and severity of floods as well. In the United States, the U.S. Army Corps of Engineers maintains a network of such flood control dams.

In areas prone to urban flooding, one solution is the repair and expansion of man-made sewer systems and stormwater infrastructure. Another strategy is to reduce impervious surfaces in streets, parking lots and buildings through natural drainage channels, porous paving, and wetlands (collectively known as green infrastructure or sustainable urban drainage systems (SUDS)). Areas identified as flood-prone can be converted into parks and playgrounds that can tolerate occasional flooding. Ordinances can be adopted to require developers to retain stormwater on site and require buildings to be elevated, protected by floodwalls and levees, or designed to withstand temporary inundation. Property owners can also invest in solutions themselves, such as re-landscaping their property to take the flow of water away from their building and installing rain barrels, sump pumps, and check valves.

Analysis of Flood Information

A series of annual maximum flow rates in a stream reach can be analyzed statistically to estimate the 100-year flood and floods of other recurrence intervals there. Similar estimates from many sites in a hydrologically similar region can be related to measurable characteristics of each drainage basin to allow indirect estimation of flood recurrence intervals for stream reaches without sufficient data for direct analysis.

Physical process models of channel reaches are generally well understood and will calculate the depth and area of inundation for given channel conditions and a specified flow rate, such as for use in floodplain mapping and flood insurance. Conversely, given the observed inundation area of a recent flood and the channel conditions, a model can calculate the flow rate. Applied to various potential channel configurations and flow rates, a reach model can contribute to selecting an optimum design for a modified channel. Various reach models are available as of 2015, either 1D models (flood levels measured in the channel) or 2D models (variable flood depths measured across the extent of a floodplain). HEC-RAS, the Hydraulic Engineering Center model, is among the most popular software, if only because it is available free of charge. Other models such as TUFLOW combine 1D and 2D components to derive flood depths across both river channels and the entire floodplain.

Physical process models of complete drainage basins are even more complex. Although many processes are well understood at a point or for a small area, others are poorly understood at all scales, and process interactions under normal or extreme climatic conditions may be unknown. Basin models typically combine land-surface process components (to estimate how much rainfall or snowmelt reaches a channel) with a series of reach models. For example, a basin model can calculate the runoff hydrograph that might result from a 100-year storm, although the recurrence interval of a storm is rarely equal to that of the associated flood. Basin models are commonly used in flood forecasting and warning, as well as in analysis of the effects of land use change and climate change.

Flood Forecasting

Anticipating floods before they occur allows for precautions to be taken and people to be warned so that they can be prepared in advance for flooding conditions. For example, farmers can remove animals from low-lying areas and utility services can put in place emergency provisions to re-route services if needed. Emergency services can also make provisions to have enough resources available ahead of time to respond to emergencies as they occur. People can evacuate areas to be flooded.

In order to make the most accurate flood forecasts for waterways, it is best to have a long time-series of historical data that relates stream flows to measured past rainfall events. Coupling this historical information with real-time knowledge about volumetric capacity in catchment areas, such as spare capacity in reservoirs, ground-water levels, and the degree of saturation of area aquifers is also needed in order to make the most acrate flood forecasts.

Radar estimates of rainfall and general weather forecasting techniques are also important components of good flood forecasting. In areas where good quality data is available, the intensity and height of a flood can be predicted with fairly good accuracy and plenty of lead time. The output of a flood forecast is typically a maximum expected water level and the likely time of its arrival at key locations along a waterway, and it also may allow for the computation of the likely statistical return period of a flood. In many developed countries, urban areas at risk of flooding are protected against a 100-year flood - that is a flood that has a probability of around 63% of occurring in any 100-year period of time.

According to the U.S. National Weather Service (NWS) Northeast River Forecast Center (RFC) in Taunton, Massachusetts, a rule of thumb for flood forecasting in urban areas is that it takes at least 1 inch (25 mm) of rainfall in around an hour's time in order to start significant ponding of water on impermeable surfaces. Many NWS RFCs routinely issue Flash Flood Guidance and Headwater Guidance, which indicate the general amount of rainfall that would need to fall in a short period of time in order to cause flash flooding or flooding on larger water basins.

In the United States, an integrated approach to real-time hydrologic computer modelling utilizes observed data from the U.S. Geological Survey (USGS), various cooperative observing networks, various automated weather sensors, the NOAA National Operational Hydrologic Remote Sensing Center (NOHRSC), various hydroelectric companies, etc. combined with quantitative precipitation forecasts (QPF) of expected rainfall and/or snow melt to generate daily or as-needed hydrologic forecasts. The NWS also cooperates with Environment Canada on hydrologic forecasts that affect both the USA and Canada, like in the area of the Saint Lawrence Seaway.

The Global Flood Monitoring System, "GFMS," a computer tool which maps flood conditions worldwide, is available online. Users anywhere in the world can use GFMS to determine when floods may occur in their area. GFMS uses precipitation data from NASA's Earth observing satellites and the Global Precipitation Measurement satellite, "GPM." Rainfall data from GPM is combined with a land surface model that incorporates vegetation cover, soil type, and terrain to determine how much water is soaking into the ground, and how much water is flowing into streamflow.

Users can view statistics for rainfall, streamflow, water depth, and flooding every 3 hours, at each 12 kilometer gridpoint on a global map. Forecasts for these parameters are 5 days into the future. Users can zoom in to see inundation maps (areas estimated to be covered with water) in 1 kilometer resolution.

Deadliest Floods

Below is a list of the deadliest floods worldwide, showing events with death tolls at or above 100,000 individuals.

Death toll	Event	Location	Date
2,500,000–3,700,000	1931 China floods	China	1931
900,000–2,000,000	1887 Yellow River (Huang He) flood	China	1887
500,000–700,000	1938 Yellow River (Huang He) flood	China	1938
231,000	Banqiao Dam failure, result of Typhoon Nina. Approximately 86,000 people died from flooding and another 145,000 died during subsequent disease.	China	1975
230,000	Indian Ocean tsunami	Indonesia	2004
145,000	1935 Yangtze river flood	China	1935
100,000+	St. Felix's Flood, storm surge	Netherlands	1530
100,000	Hanoi and Red River Delta flood	North Vietnam	1971
100,000	1911 Yangtze river flood	China	1911

In Myth and Religion

Flood myths (great, civilization-destroying floods) are widespread in many cultures.

Flood events in the form of divine retribution have also been described in religious text. As a prime example, the Genesis flood narrative plays a prominent role in Judaism, Christianity and Islam.

Flood Stage

Flood stage is the level at which a body of water's surface has risen to a sufficient level to cause sufficient inundation of areas that are not normally covered by water, causing an inconvenience or a threat to life and/or property. When a body of water rises to this level, it is considered a flood event. Flood stage does not apply to areal flooding. Because areal flooding occurs, by definition, over areas not normally covered by water, any water at all creates a flood. Usually, Moderate and Major stages are not defined for areal floodplains.

Definition

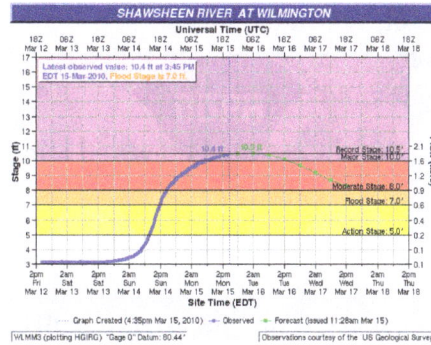

Example graph of stream stages showing Action Stage, Flood Stage, Moderate Stage, Major Stage, and Record Stage on a river.

Flood stage is the water level, as read by a stream gauge or tide gauge, for a body of water at a particular location, measured from the level at which a body of water threatens lives, property, commerce, or travel. The term "at flood stage" is commonly used to describe the point at which this occurs. "Gauge height" (also referred to as "stream stage", "stage of the [body of water]", or simply "stage") is the level of the water surface above an established zero datum at a given location. The zero level can be arbitrary, but it is usually close to the bottom of the stream or river or at the average level of standing bodies of water. Stage was traditionally measured visually using a staff gauge, which is a fixed ruler marked in 1/100 and 1/10 foot intervals, however electronic sensors that transmit real-time information to the Internet are now used for many of these kind of measurements. The flood stage measurements are given as a height above or below the zero level. Levels below zero are reported as a negative value.

While usually the flood stage is set at the elevation of the floodplain, it can be higher (if there are no structures, roads, or farming areas immediately on the floodplain) or lower (if there are structures such as marinas, lake houses, or docks low on the banks or shores of the body of water) depending on the location. Because flood stage is defined by impacts to people, as opposed to the natural topography of the area, flood stages are usually only calculated for bodies of water near communities.

The flood stage can be listed for an entire community, in which case it is often set to the lowest man-made structure or road in the area, the lowest farming field in the area, or the floodplain. It can also be set for a specific location ("flood stage is 12 feet on Maple Street at First Avenue" means that the specified intersection will begin to flood when the stage reaches 12 feet (3.7 m)).

In the United States during flood events, the National Weather Service will issue flood warnings that list the current and predicted stages for affected communities as well as the local flood stage. Current stage data is collected by the USGS using a network of gauges, over 9000 of which transmit real time data via satellite, radio, or telephone. Many communities have inundation maps that provide information on which areas will flood at which stages.

Flood Categories

In the United States, there are five levels of flooding.

Action Stage

- Rivers: typically at this level, the water surface is generally near or slightly above the top of its banks, but no man-made structures are flooded; typically any water overflowing is limited to small areas of parkland or marshland.

- Coastlines: at action stage, usually elevated tides and minor inundation of low-lying beach areas occurs.

Minor Flood Stage

- Rivers: minor flooding is expected at this level, slightly above flood stage. Few, if any, buildings are expected to be inundated, however, roads may be covered with water, parklands, and lawns may be inundated and water may go under buildings on stilts or higher elevations.

- Coastlines: water will usually run all the way up to the dune in waves during a minor flood. Overwash may occur on shoreline roads. Lifeguard structures and beach concession stands will usually be flooded and may be damaged by surf.

Moderate Flood Stage

- Rivers: inundation of buildings usually begins at this stage. Roads are likely to be closed and some areas cut off. Some evacuations may be necessary.

- Coastlines: at moderate flood stage, usually water overtops the natural dune and begins flooding coastal areas. Shoreline roadways and beaches will often be completely flooded out. High surf usually associated with this level of flooding may pound some oceanside structures like piers, boardwalks, docks, and lifeguard stations apart. Beach houses may be damaged by water and surf, especially if lacking stilts.

Major Flood Stage

- Rivers: significant to catastrophic, life-threatening flooding is usually expected at this stage. Extensive flooding with some low-lying areas completely inundated is likely. Structures may be completely submerged. Large-scale evacuations may be necessary.

- Coastlines: Water surges over not only the dune, but also man-made walls and roads. Large and destructive waves pound weak structures to bits and severely damage well-built homes and businesses. Overwash occurs on high-level seawalls. If major flooding occurs at high tide, impacts may be felt well inland. If cities are at or below sea level, catastrophic flooding can inundate the entire city and cause millions or billions of dollars in damage (such as occurred in New Orleans during Hurricane Katrina).

Record Flood Stage

- Rivers: at this level, the river is at its highest that it has been since records began for the area where the stream gauge is located. This does not necessarily imply a major flood. Some areas may have never experienced major flooding, and thus record stage is in the moderate category.

- Coastlines: Usually, record flooding at the coast is associated with tropical cyclones, but it may be associated with coastal storms, Nor'easters, seiches caused by earthquakes, strong thunderstorms, or tsunamis. Destruction is often extensive and may extend a far distance inland.

Flood Warning

Flood warning is closely linked to the task of flood forecasting. The distinction between the two is that the outcome of flood forecasting is a set of forecast time-profiles of channel flows or river levels at various locations, while "flood warning" is the task of making use of these forecasts to make decisions about whether warnings of floods should be issued to the general public or whether previous warnings should be rescinded or retracted.

Description

The task of providing warning for floods is divided into two parts:

- decisions to escalate or change the state of alertness internal to the flood warning service provider, where this may sometimes include partner organisations involved in emergency response;

- decisions to issue flood warnings to the general public.

The decisions made by someone responsible for initiating flood warnings must be influenced by a number of factors, which include:

- The reliability of the available forecasts and how this changes with lead-time.

- The amount of time that the public would need to respond effectively to a warning.

- The delay between a warning being initiated and it being received by the public.

- The need to avoid issuing warnings unnecessarily, because of the wasted efforts of those who respond and because a record of false alarms means that fewer would respond to future warnings.

- The need to avoid situations where a warning condition is rescinded only for the warning to be re-issued within a short time, again because of the wasted efforts of the general public and because such occurrences would bring the flood warning service into disrepute.

A computer system for flood warning will usually contain sub-systems for:

- flood forecasting;

- automatic alerting of internal staff;

- tracking of alert messages and acknowledgements received;

- diversion of messages to alternates where no acknowledgement received.

National Flood Warning Services

The type of flood warning service available varies greatly from country to country, and a location may receive warnings from more than one service.

United Kingdom

Arrangements for flood warnings vary across the United Kingdom with several agencies leading on warnings for emergency responders and the public. The Environment Agency, Natural Resources Wales and Scottish Environment Protection Agency all undertake location specific flood warning activities for communities at risk depending upon the scale of flood risk, technical challenges and investment needed to deliver a reliable service.

Prior to issuing a flood warning consideration is given to:

- the needs of communities to activate emergency response plans
- the nature of the catchment or coastline and the lead time that may be provided
- meteorological observations and forecast information on rainfall and coastal water levels
- hydrological observations and flood forecasts
- reference to thresholds of historic or forecast flood levels

Dissemination of flood warnings has moved towards a service whereby those at risk can pre-register to receive warnings by phone, email or text message from an automatic system, Floodline. Both warnings and updates about current conditions are also carried by local radio stations. In addition, live updates are carried by the Environment Agency's website, showing which locations have flood warnings in place and the severity of these warnings.

There is currently no flood warning system in Northern Ireland, but the Met Office does issue weather warnings. Flood risk management is the responsibility of Rivers Agency in Northern Ireland. Consideration will be given to the introduction of a warning system as part of the implementation of the EU Floods directive.

United States

In the United States, the National Weather Service issues flood watches and warnings for large-scale, gradual river flooding. Watches are issued when flooding is possible or expected within 12–48 hours, and warnings are issued when flooding over a large area or river flooding is imminent or occurring. Both can be issued on a county-by-county basis or for specific rivers or points along a river. When rapid flooding from heavy rain or a dam failure is expected, flash flood watches and warnings are issued.

In the U.S. and Canada, dissemination of flood warnings is covered by Specific Area Message Encoding (SAME) code FLW, which is used by the U.S. Emergency Alert System and NOAA Weather Radio network and in Canada's Weatheradio Canada network.

"Flood statements" are issued by the National Weather Service to inform the public of flooding along major streams in which there is not a serious threat to life or property. They may also follow a flood warning to give later information.

Example of a Flood Warning

The following is an example of a "Flood Warning." The South Chickamauga Creek is used as an example:

```
608

WGUS44 KMRX 210433 CCA

FLWMRX


BULLETIN - IMMEDIATE BROADCAST REQUESTED

FLOOD WARNING

NATIONAL WEATHER SERVICE MORRISTOWN, TN

1233 AM EDT MON SEP 21 2009


...THE NATIONAL WEATHER SERVICE IN MORRISTOWN, TN HAS ISSUED A FLOOD

WARNING FOR THE FOLLOWING RIVERS SOUTH CHICKAMAUGA IN GEORGIA...

TENNESSEE...


   SOUTH CHICKAMAUGA CREEK NEAR CHATTANOOGA TN AFFECTING CATOOSA AND

HAMILTON COUNTIES


HEAVY RAINFALL SUNDAY EVENING OF AROUND 3 INCHES IN THE CHATTANOOGA

AREA HAS CAUSED THE SOUTH CHICKAMAUGA CREEK TO RISE RAPIDLY.


GAC047-TNC065-211821-

/O.NEW.KMRX.FL.W.0013.090921T0433Z-090922T2200Z/

/CHKT1.1.ER.090921T0421Z.090922T0600Z.090922T1600Z.NO/

1233 AM EDT MON SEP 21 2009
```

THE NATIONAL WEATHER SERVICE IN MORRISTOWN, TN HAS ISSUED A

* FLOOD WARNING FOR

  THE SOUTH CHICKAMAUGA CREEK AT CHICKAMAUGA TN
* FROM THIS MORNING TO TUESDAY EVENING.
* AT 11:15 PM SUNDAY EVENING THE STAGE WAS 13.0 FEET.
* MINOR FLOODING IS FORECAST.
* FORECAST...THE RIVER WILL RISE ABOVE FLOOD STAGE AROUND 7 AM MONDAY

  AND CREST NEAR 19.0 FEET AROUND 2 AM TUESDAY. THE RIVER WILL FALL

  BELOW FLOOD STAGE TUESDAY AFTERNOON.
* AT 19.0 FEET...WATER ACROSS MACK SMITH ROAD BEGINS TO IMPEDE

  TRAFFIC. WEST CHICKAMAUGA CREEK OVERFLOWS ITS BANKS NEAR THE

  GEORGIA STATE LINE AND INUNDATES SEVERAL ROADS AND PROPERTIES IN

  THE AREA.

$$

PRECAUTIONARY/PREPAREDNESS ACTIONS...

MOST DEATHS IN FLOODS OCCUR IN CARS! IF YOU COME TO A CLOSED OR

FLOODED ROAD, TURN AROUND! DON'T DROWN! FOR MORE DETAILS, STAY TUNED

TO NOAA WEATHER RADIO OR COMMERCIAL TELEVISION OR RADIO THAT CARRY

WEATHER INFORMATION.

& &

$$

TD

Example of a Flood Statement

000

WGUS84 KMRX 221008

FLSMRX

FLOOD STATEMENT

NATIONAL WEATHER SERVICE MORRISTOWN, TN

608 AM EDT TUE SEP 22 2009

...THE FLOOD WARNING CONTINUES FOR THE FOLLOWING RIVERS IN GEORGIA...

TENNESSEE...

  SOUTH CHICKAMAUGA CREEK @ CHICKAMAUGA TN AFFECTING CATOOSA AND

HAMILTON COUNTIES

IN HAMILTON COUNTY...THERE ARE TOO MANY STREETS CLOSED TO MENTION.

IN SHORT...BE PREPARED FOR LONG DELAYS. ALSO...FAR TOO MANY PEOPLE

ARE TRYING TO DRIVE THROUGH FLOOD WATERS. THIS IS A GOOD WAY TO DIE.

PRECAUTIONARY/PREPAREDNESS ACTIONS...

MOST DEATHS IN FLOODS OCCUR IN CARS! IF YOU COME TO A CLOSED OR

FLOODED ROAD, TURN AROUND! DON'T DROWN! DRIVING INTO FLOODED ROADS

IS A GOOD WAY TO DIE. BETTER TO BE LATE THAN END UP ON THE NEWS.

&&

GAC047-TNC065-221608-

```
/O.CON.KMRX.FL.W.0013.000000T0000Z-090923T1942Z/

/CHKT1.2.ER.090921T0923Z.090922T1800Z.090923T1342Z.NO/

608 AM EDT TUE SEP 22 2009


THE FLOOD WARNING CONTINUES FOR

  THE SOUTH CHICKAMAUGA CREEK @ CHICKAMAUGA TN

* UNTIL WEDNESDAY AFTERNOON.

* AT  5:15 AM TUESDAY THE STAGE WAS 25.8 FEET.

* MODERATE FLOODING IS OCCURRING. THE FORECAST IS FOR MODERATE

  FLOODING TO CONTINUE.

* FORECAST...THE RIVER WILL CONTINUE TO RISE AND CREST NEAR 26.5 FEET

  TUESDAY AFTERNOON. THE RIVER WILL FALL BELOW FLOOD STAGE LATE

  WEDNESDAY MORNING.

* AT 27.0 FEET...HOMES ON ARLENA CIRCLE (OFF SHALLOWFORD ROAD) ARE

  EVACUATED.  EVACUATIONS ALSO TAKE PLACE AT THE FOUNTAINBLEAU

  APARTMENTS ON SPRING CREEK ROAD IN EAST RIDGE.

$$

$$

BOYD
```

Iowa Flood Center

The Iowa Flood Center at the University of Iowa operates the largest real-time flood monitoring system of its kind in the world. It includes more than 200 real-time stream stage sensors that feed data into the Iowa Flood Information System where data can be viewed, online, by disaster management staff and the general public. The stream stage sensors, mounted on bridges and culverts, use ultrasonic sensors to monitor stream and river levels.

Floodgate

Floodgates, also called stop gates, are adjustable gates used to control water flow in flood barriers, reservoir, river, stream, or levee systems. They may be designed to set spillway crest heights in dams, to adjust flow rates in sluices and canals, or they may be designed to stop water flow entirely as part of a levee or storm surge system. Since most of these devices operate by controlling the water surface elevation being stored or routed, they are also known as crest gates. In the case of flood bypass systems, floodgates sometimes are also used to lower the water levels in a main river or canal channels by allowing more water to flow into a flood bypass or detention basin when the main river or canal is approaching a flood stage.

Tokyo floodgates created to protect from typhoon surges

Types

Bulkhead gates are vertical walls with movable, or re-movable, sections. Movable sections can be lifted to allow water to pass underneath (as in a sluice gate) and over the top of the structure. Historically, these gates used stacked timbers known as stoplogs or wooden panels known as flashboards to set the dam's crest height. Some floodgates known as coupures in large levee systems slide sideways to open for various traffic. Bulkhead gates can also be made of other materials and used as a single bulkhead unit. Miter gates are used in ship locks and usually close at an 18° angle to approximate an arch.

A sluice gate on the Harran canal

A flood wall gate at Harlan, Kentucky

Hinged crest gates, are wall sections that rotate from vertical to horizontal, thereby varying the height of the dam. They are generally controlled with hydraulic power, although some are passive and are powered by the water being impounded. Variations:

- flap gate
- fish-belly flap gates
- Bascule gates
- Pelican gates

A hinged crest gate during installation

Fish belly flap gates at the Scrivener Dam, Canberra

Radial gates are rotary gates consisting of cylindrical sections. They may rotate vertically or horizontally. Tainter gates are a vertical design that rotates up to allow water to pass underneath. Low friction trunnion bearings, along with a face shape that balances hydrostatic forces, allow this design to close under its own weight as a safety feature.

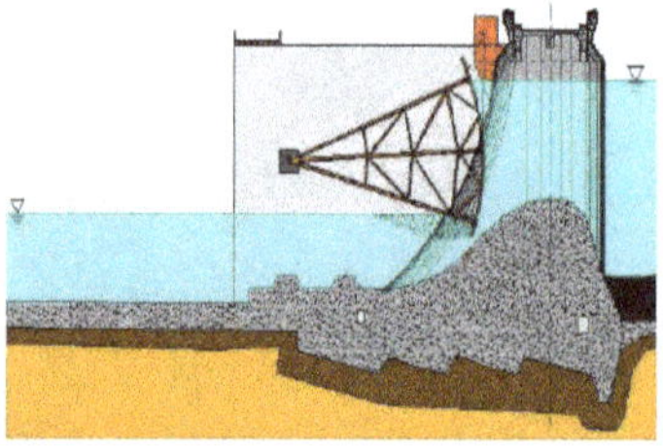

Tainter gate diagram

Tainter gates and spillway

Drum gates are hollow gate sections that float on water. They are pinned to rotate up or down. Water is allowed into or out of the flotation chamber to adjust the dam's crest height.

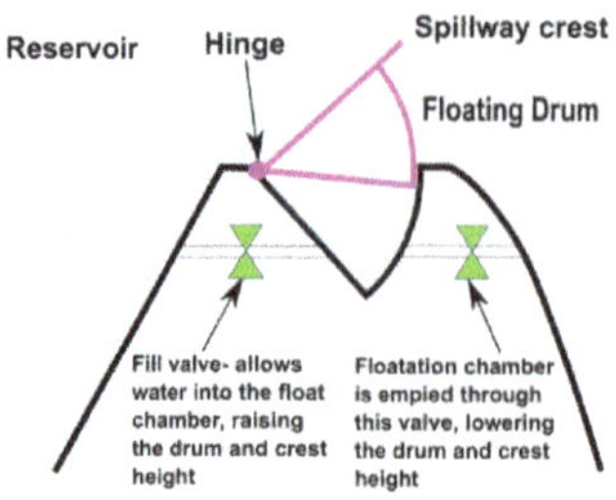

Drum gates are controlled with valves.

Drum gates on a diversion dam

- **Roller gates** are large cylinders that move in an angled slot. They are hoisted with a chain and have a cogged design that interfaces with their slot.

- **Clamshell gates** have an external clamshell leaf design.

A roller gate on the Mississippi.

Clamshell floodgates at the Arrowrock Dam.

Fusegates are a mechanism designed to provide the controlled release of water in the event of exceptionally large floods. The design consists of free standing blocks (the fusegates) set side by side on a flattened spillway sill. The Fusegate blocks act as a fixed weir most of the time, but in excessive flood conditions they are designed to topple forward, allowing the controlled discharge of water. Multiple fusegates are generally set up side by side, with each fusegate designed to release under progressively extreme flooding, thus minimizing the impact of the floodwater on the river downstream. The System is developed and patented by Hydroplus from Paris, France. It has been installed on more than 50 dams around the world with sizes ranging from 1 m to more than 9 m in height. Fusegate are typically used to increase the storage capacity of existing dams or to maximize the discharge potential of undersized spillways.	Typical fusegate sketch	Fusegate in Terminus Dam - Lake Kaweah
Mitre gates		

Valves

Discharge from a Howell-Bunger valve

Valves used in floodgate applications have a variety of design requirements and are usually located at the base of dams. Often, the most important requirement (besides regulating flow) is energy dissipation. Since water is very heavy, it exits the base of a dam with the enormous force of water pushing from above. Unless this energy is dissipated, the flow can erode nearby rock and soil and damage structures.

Other design requirements include taking into account pressure head operation, the flow rate, whether the valve operates above or below water, and the regulation of precision and cost.

- Fixed cone valves are designed to dissipate the energy from a water flow during reservoir discharge. They are a round pipe section with an adjustable sleeve gate and cone at the discharge end. Flow is varied by moving the sleeve away or towards its cone seat. The design allows high pressure water from the base of a dam to be released without causing erosion to the surrounding environment. Fixed cone valves are able to handle heads up to 300 m.

- Hollow jet valves are a type of needle valve used for floodgate discharge. A cone and seat are inside a pipe. Water flows through an annular gap between the pipe and cone when it is moved downstream, away from the seat. Ribs support the bulb assembly and supply air for water jet stabilization.

- Ring jet valves are similar to fixed cone valves, but have an integral collar that discharges water in a narrow stream. They are suitable for heads up to 50 m.

- Jet flow gate, similar to a gate valve but with a conical restriction prior to the gate leaf that focuses the water into a jet. They were developed in the 1940s by the United States Bureau of Reclamation to allow fine control of discharge flow without the cavitation seen in regular gate valves. Jet flow gates are able to handle heads up to 150 m.

Physics

In order to do a simple calculation of the force on a rectangular flood gate one can use the following equation:

$$F = pA$$

where:

F = force measured in the SI units kg·m·s^{-2} which is called the newton (N)

p = pressure = ρgh measured in N/m², which is called the pascal (Pa)

where:

- ρ (rho) is the density of fresh water (1000 kg/m³);

- g is the acceleration due to gravity on Earth (9.8 m/s²);

- h is the height of the water column in meters.

A = area = rectangle : length × height measured in m²

where:

length = the horizontal length of a rectangular floodgate measured in meters

height = the height of a non-submerged flood gate from the bottom of the water column to the water surface measured in meters

If the rectangular flood gate is submerged below the surface the same equation can be used but only the height from the water surface to the middle of the gate must be used to calculate the force on the flood gate.

Flood Barrier

The Oosterscheldekering contains 62 steel doors, each 42 metres (138 ft) wide

A flood barrier, surge barrier or storm surge barrier is a specific type of floodgate, designed to prevent a storm surge or spring tide from flooding the protected area behind the barrier. A surge barrier is almost always part of a larger flood protection system consisting of floodwalls, levees (also known as dikes), and other constructions and natural geographical features.

Flood barrier may also refer to barriers placed around or at individual buildings to keep floodwaters from entering those buildings.

The Maeslantkering closes the main entrance to the Port of Rotterdam, the largest port in Europe.

Flood Barriers Around The World

Delta Works

The Delta Works in the Netherlands is the largest flood protection project in the world. This project consists of a number of surge barriers, the Oosterscheldekering being the largest surge barrier in the world, 9 kilometres (5.6 mi) long. Other examples include the Maeslantkering, Haringvliet-dam and the Hartelkering.

River Thames Flood Barrier

Thames Barrier

The Thames Barrier is the world's second largest movable flood barrier (after the Oosterschelde-kering and the Haringvlietdam) and is located downstream of central London. Its purpose is to prevent London from being flooded by exceptionally high tides and storm surges moving up from the North Sea. It needs to be raised (closed) only during high tide; at ebb tide it can be lowered to release the water that backs up behind it.

The IHNC Surge Barrier, being built by the US Army Corps of Engineers. The GIWW in the foreground, the MRGO in the background

New Orleans

In 2009 the United States Army Corps of Engineers started construction of an ambitious project that aimed to prevent storm surges from flooding the city by 2011. The IHNC Lake Borgne Surge Barrier on the confluence of these waterways is the largest in the United States. The new Seabrook floodgate prevents a storm surge from entering from Lake Ponchartrain. The GIWW West Closure Complex closes the Gulf Intracoastal Waterway to protect the west side of the city. This complex is unique in that it contains the world's largest pumping station, necessary to pump out rainwater that is discharged in the protected side of the canal during a hurricane.

Eider Barrage, landward side, open

Eider Barrage

The Eider Barrage is located at the mouth of the river Eider near Tönning on Germany's North Sea coast. Its main purpose is protection from storm surges by the North Seas. It is Germany's largest coastal protection structure.

St. Petersburg Dam

The Saint Petersburg Dam (officially called the Saint Petersburg Flood Prevention Facility Complex) is a 16 km (9.9 mi) barrier separating the Gulf of Finland from Neva Bay to protect the city of Saint Petersburg, Russia from coastal flooding. The Soviet Union started construction of the barrier in 1978 and it was completed and made operational in 2011.

New England

The New Bedford Harbor Hurricane Barrier protects the city of New Bedford, Massachusetts, with a mostly immovable barrier of stone and fill. It has three land and one marine door for access in calm seas.

The nearby Fox Point Hurricane Barrier protects the city of Providence, Rhode Island.

The US Army Corps of Engineers also owns and operates the hurricane barrier at Stamford, CT.

Venice

The MOSE Project is intended to protect the city of Venice, Italy, and the Venetian Lagoon from flooding.

River Foss Barrier

The river Foss, York, UK has a barrier to control the inflow of fast moving water from the river ouse that may overspill its banks upstream the foss and flood surrounding properties. Animations and photos explain it.

Local Flood Barriers

Flood barriers may be placed temporarily or permanently around individual buildings or at building entrances to keep floodwaters from entering those buildings. A wall constructed of sandbags is an example of a temporary barrier. A reinforced concrete wall is an example of a permanent barrier.

Flood barriers can be manufactured to meet governmental or industry standards. Certification is available through third party testing laboratories.

The Water-Gate Flood barrier is a rapid response barrier which can be rolled out in minutes. It is unique in the way that it self deploys using the weight of water to hold it back. The product has been FM Approved following testing from the US Army. It is used in 30 countries around the world, and notably by the Environment Agency in the UK.

Permissions

Index

www.ingramcontent.com/pod-product-compliance
Lightning Source LLC
Chambersburg PA
CBHW061313190326
41458CB00011B/3796